Ben Miller is, like you, a mutant ape living through an Ice Age on a ball of molten iron, circulating a supermassive black hole. He is also an actor, comedian and approximately one half of Armstrong & Miller. He's a presenter of ITV's *It's Not Rocket Science*, and plays Professor McTaggart, a disembodied brain, in *Horrible Science* on CiTV. He continues to struggle with the idea that he may never be an astronaut.

@ActualBenMiller

Praise for Ben Miller

'A witty romp through the cosmos ... Miller provides a light-hearted but thorough exploration of the science behind the search for intelligent alien life – what it might be like, where and how we might find it – and reflects on whether we could understand or recognize an alien at all'
Science

'Whether it's an objective look at UFO encounters, detailing the challenges of contacting aliens or explaining how life on Earth can inform our search, his snappy, conversational style will keep you turning the pages'
Discovery

'Miller has a gift for making complex ideas seem comprehensible ... enjoyable and educational ... full marks, Miller, for a tricky job well done'
Mail on Sunday

'Books like these should act as gateway drugs for the incurably curious'
Observer

'*It's Not Rocket Science* carries you along in a veritable tsunami of fascinating facts and eureka moments'
Daily Mail

'*It's Not Rocket Science* is a wonderful handbook for the curious ... it's so entertainingly written that the learning doesn't hurt at all ... could be the best bit of edutainment you read this year'
Irish Times

Also by Ben Miller

It's Not Rocket Science

THE ALIENS ARE COMING!

The Exciting and Extraordinary Science Behind
Our Search for Life in the Universe

Ben Miller

sphere

SPHERE

First published in Great Britain in 2016 by Sphere
This paperback edition published in 2017 by Sphere

A CIP catalogue record for this book
is available from the British Library.

ISBN 978-0-7515-4504-3

Typeset in Bembo by M Rules
Printed and bound in Great Britain by
Clays Ltd, St Ives plc

Papers used by Sphere are from well-managed forests
and other responsible sources.

Sphere
An imprint of
Little, Brown Book Group
Carmelite House
50 Victoria Embankment
London EC4Y 0DZ

An Hachette UK Company
www.hachette.co.uk

www.littlebrown.co.uk

For Sonny, Harrison and Lana

CONTENTS

CHAPTER ONE

Extremophiles

In which the author gets his starships in a row, and discovers that the solar system is an oasis, not a desert.

WE COME IN PEACE

On 25 August 2012, the first of our ships reached interstellar space. It was unmanned. Launched three and a half decades earlier, it had skirted Jupiter and Saturn, and was now heading out of the solar system towards Camelopardalis, a little-known constellation close to the Plough. Although clear of the solar wind, it was not quite out of reach of the sun's gravity, nor would it be for a further thirty millennia. By then it would finally have traversed what is known as the Oort Cloud, a thick outer shell of icy rubble that encases our home star and its eight planets like the flesh around a peach stone. At that point it would be nearly a light year out. Forget the galaxy; even the solar system is unimaginably large.

The ship's name was *Voyager 1*, and on board was a message from the people of Earth, encoded on what became known as the 'Golden Record'. This gold-plated phonographic disc, curated by the distinguished American cosmologist Carl Sagan, spoke on behalf of all humanity. It began with a recorded message from the Secretary-General of the United Nations, Kurt Waldheim. Reading haltingly, with a strong Austrian accent, he made the following statement:

> I send greetings on behalf of the people of our planet. We step out of our solar system into the universe seeking only peace and friendship: to teach if we are called upon; to be taught if we are fortunate. We know full well that our planet and all its inhabitants are but a small part of this immense universe that surrounds us and it is with humility and hope that we take this step.

After this greeting came a choir of voices speaking in fifty-five languages:[1] everything from Akkadian, the language of ancient Sumer, to Wu, the contemporary Chinese dialect spoken around Shanghai. Some, such as the Japanese, appeared shy: 'Hello, how are you?' Others were more forthcoming, such as the Amoy of south-eastern China, who offered: 'Friends of space, how are you all? Have you eaten yet? Come visit us if you have time.' The speaker of Ancient Greek, on the other hand, issued a barely concealed threat: 'Greetings to you all, whoever you are. We come in friendship . . . to those who are friends.'

These Greetings of Earth were accompanied by twenty-odd Sounds of Earth: among them echoing footsteps, hard

---

1 For this, as in all matters Voyager, I refer to the peerless *Haynes Manual NASA Voyager 1 & 2* by Christopher Riley, Richard Corfield and Philip Dolling.

rain and a handsaw cutting fresh wood. Over one hundred Scenes of Earth showed images such as a hand being X-rayed, the chemical structure of DNA and a man and a pregnant woman in silhouette. And, finally, there was the Music of Earth, with over twenty of humanity's finest recordings, including the first movement of Beethoven's Fifth, Bach's *Well-Tempered Clavier*, and 'Johnny B. Goode' performed by Chuck Berry.

EARTH LEFT YOU A MESSAGE

That might have the flavour of science fiction – at least, I hope it does as I was trying my hardest – but it is all true. So far as we know, the Golden Record has not yet been intercepted by spacefaring aliens. If it is, what they are to make of it is anyone's guess. For a start, we have to hope that they aren't too big. An alien the size of a blue whale might have a hard time getting a needle in the groove, let alone building a hi-fi system for it to play on at the required speed of 16⅔ revolutions per minute. Equally, too small an alien – one the size of a microbe, say – might never realise the Golden Record, or *Voyager 1* itself, was even there in the first place.

Next, of course, we have to hope that they share our perception of time. As we shall see in a later chapter, not all animals on Earth do, let alone all aliens. To crows, for example, whose brains have a faster clock, human communication appears slow and deliberate. If 'alien time' passes much faster than 'human time', the aliens might not realise that human speech contains information; it might simply sound like long, unintelligible groans. To understand human speech, it helps to have a brain that chugs along at human speed.

And while we are on the subject of human speech, we had better hope that any aliens that find the Golden Record have ears, that the frequency range of those ears matches that of our own, and that they themselves communicate using vocalisations. On a deeper level, we had better hope that the concepts expressed in our messages – things like 'peace' and 'space' and 'time' – have equivalents within their own language, or languages. And on an even deeper level, we hope they share the concept of a 'message' within their culture, and don't just fire it straight back again.

It doesn't end there. To be able to see the instructions on the case of the Golden Record, detailing how the information inside is to be decoded, the aliens had better be able to see, and their vision had better be attuned to the same range of the electromagnetic spectrum as our own eyes. Again we can see from life forms on Earth that this is not a given. A race of superintelligent bats, for example, might see the Golden Record as nothing more than a metallic Frisbee. A community of superintelligent bacteria might just see it as a snack.

And, most importantly of all, we had better hope that the aliens have a good understanding of human culture. If they don't, they are going to have a hard time figuring out what we were up to. When storage capacity must have been so precious, why include so many greetings? Why are genitals shown in some of the drawings of humans, but not in others? What's the music for? Who are these people, and what the hell are they trying to tell us?

In short, we had better hope that the aliens are just like us.

LOVING THE ALIEN

Is there any question more fascinating than whether or not we are alone in the universe? The faint, ghostly light of the Milky Way is the glow of billions of stars. Is it really possible that Earth is the only habitable planet among them, and that we are the only intelligent species? And if there is intelligent life out there, might we be able to communicate with it?

The Ancient Greeks certainly thought we might. Epicurus, for example, one of the founding fathers of modern science, stated around 300 BC that 'other worlds, with plants and other living things, some of them similar and some of them different from ours, must exist'. Newton was also onside, as is plain from an appendix he added to his famous treatise on mechanics and gravitation, the *Principia*:

> This most beautiful System of the Sun, Planets, and Comets, could only proceed from the counsel and dominion of an intelligent and powerful being. And if the fixed Stars are the centers of other like systems, these, being form'd by the like wise counsel, must be all subject to the dominion of One.

Aliens are everywhere. They can be angels come to warn us of the follies of nuclear war or they can be demons that abduct us to carry out bizarre sexual experiments. Their shape changes, from the angry Little Green Men of the first half of the twentieth century to the placid Greys of the present day. They visit us in flying saucers, speak to us telepathically, or appear as strange lights in the sky. Yet so far as we can determine, all this is a product of our imaginations. Much as we might wish it were otherwise, there is no compelling evidence that intelligent, technologically advanced aliens have ever visited Earth.

But before you throw this book down in a pique of anti-scientific disgust and head for the Mind, Body and Spirit section, stop. Because as is so often the case, the real science is so much more interesting than the non-scientific stuff. While alien autopsies grab the headlines, thousands of scientists – real, hardworking, peer-reviewed, genuinely qualified scientists – are slowly inching closer to the real thing. And trust me: if we do manage to make contact with an alien intelligence, those stories about flying saucers and pervy Little Green Men are going to seem very man-made indeed.

The plain truth is that the last few years have seen something of a sea change in the way we view life in the cosmos. Thanks to NASA's recent Kepler mission, we have discovered that planets like ours are common throughout the galaxy. We also know that life got started on Earth very early in its history, and that it thrives in some incredibly extreme environments. As our probes and manned missions venture out into the solar system, and we image Earth-like planets with ever-increasing accuracy, our first encounter with alien life is rapidly approaching.

Most scientists expect that encounter will take place via a telescope, and that the life in question will be in the form of single-celled organisms so small that they would be invisible to the naked eye. A second, slightly more remote possibility is that microscopic organisms will be found on an icy moon within our own solar system, or even living cheek by jowl with us right here on Earth. And if single-celled life is as widespread as we currently believe, complex intelligent life won't be far behind.[2] Just how far behind is the subject of this book.

---

2 Complex in the sense of being made up of connected parts. In complex life, cells are grouped into tissues which are in turn grouped into organs.

Thrillingly, it turns out that life on Earth can teach us a surprising amount about life on other planets. Complex life, as we shall see, is rarer than single-celled life; exactly how much rarer is a subject of an intense but increasingly well-informed debate. Intelligence, as we shall shortly discover, is not unique to humans; in fact we share it with at least half a dozen other species, and maybe more. Some of those other intelligent species even have language, and decoding it may be an important first step towards communicating with extraterrestrials.

Forget science fiction. You are living through one of the most extraordinary revolutions in the history of science; the emergent belief of a generation of physicists, biologists and chemists that we are not alone. Our journey to understand how this revolution has come about will lead us through some ravishingly beautiful science, and hint at answers to some truly deep existential questions. All of what follows is accessible if you have an open mind; in fact, a creative bent will be as valuable as a scientific one, because this subject goes right to the heart of what it means to be human.

Before we get started, here's the briefest of guides to the journey ahead. These three opening chapters will give us an overview of the hunt for extraterrestrials to date, UFO crazes included, and try to answer the question of why the Search for Extra-Terrestrial Intelligence, or SETI as it is known, has gone from pariah to pontiff in less than a decade. In the meat of the book, we'll get a handle on what our latest studies of life on Earth can tell us about the possibilities for 'life-as-we-know-it' and 'life-as-we-don't': in other words, the chances of finding intelligent extraterrestrial organisms that are based on carbon, and those made of something else entirely. Finally, we'll look at how we might decode an alien message, should we be lucky enough to receive one, and what kind of messages – if any – we should be sending in return.

If our science is right, within the next decade we will have hard evidence that there are other living things out there in the universe. As we shall see, it's an outside bet, but if we are very lucky, some of those living things will have been just at the right stage of development at just the right time to have sent us a message that we are capable of understanding. Some of those messages might be travelling through you right now, as you read this book. If you are at all interested in how we might intercept them, and what they might say, read on . . .

CALLING OCCUPANTS OF INTERPLANETARY CRAFT

As a child of the Space Age, I have always been fascinated by the idea of life beyond Earth. Born in 1966, I was three and a half when *Apollo 11* landed on the Moon, and although I was too small to stay up and watch the live broadcast, I clearly remember the bulletins that swamped the news the following day. Even now, as I watch the footage of Neil Armstrong stepping down from the lunar module, I feel the same exquisite mix of elation and disappointment. Elation, of course, because such an extraordinary thing is possible. And disappointment that an enormous multicoloured ten-tacle didn't reach out from behind a rock and give him a high five.

There was enough novelty in the Moon landings for my contemporaries and me to overlook the absence of aliens; the bizarre effects of low gravity and no atmosphere were more than enough to be going on with. Looking back, it's almost comical how little science is going on in those first few Apollo missions. If you ever doubt that humans are descended from chimps, just watch a few weightless astronauts turning somersaults and attempting to have a tea party as they while away the hours on their three-day journey. To me now, all that hyperactivity seems like an attempt to distract the watching billions from one disquieting central fact: the Moon is about as dead as it is possible to be.

It didn't help that kids of my age had high expectations that we might meet aliens within our lifetime. For a start, we had inherited a vault of alien–invasion Golden Age science fiction, such as Ray Bradbury's *The Martian Chronicles* and John Wyndham's *The Kraken Wakes*. Most often in these stories, aliens were out there in the darkness, watching. Mankind had ascended the throne of Technology, and it was high time we were usurped. The conventional wisdom is that these chilling stories were a manifestation of the Cold War and the threat of Soviet attack, but if you ask me there was another equally important inspiration: the birth of broadcasting.

Radio had come first, with Marconi claiming the first transatlantic radio communication in 1901, and international broadcasts playing an important part in Germany's propaganda machine in the run-up to the Second World War. Television followed soon after, dominating by the late 1940s. Both media were blasted out by giant transmitters that sent just as much signal out into the cosmos as they did to the horizon. By 1950, when Ray Bradbury published *The Martian Chronicles*, somewhere lodged within the collective unconscious was the idea that if there were technologically advanced aliens on our neighbouring planets, they knew exactly where we were and what we were up to.

Both radio and television signals, of course, are carried

by electromagnetic waves, which travel at the speed of light.[3] That leaves us with a disquieting thought. Those signals, and all of those broadcast since, have been billowing away from Earth for the best part of seventy years. There are now hundreds of star systems within range of our TV and radio signals, not just the dozen or so that would have been in range in 1950. Maybe aliens are on their way right now, enraged by the unsatisfactory ending of *Twin Peaks*.[4]

For me, this story was told best by Carl Sagan. At the beginning of *Contact*, the 1997 movie based on his book of the same name, the camera surfs the spreading wave of talk shows, news bulletins and popular music as it leaves Earth and makes its way out into the galaxy. As we gather pace, we catch up with earlier and earlier broadcasts. To begin with, we hear thrash metal and the Spice Girls. Further out, we pass Madonna, then the theme from the first *Star Wars* movie, then, further on still, we overtake Neil Armstrong's 'giant leap for mankind'. Finally, with the Milky Way Galaxy receding into the distance, we hear the announcer of *The Maxwell House Good News of 1939*, then Morse code, then silence.

FROM RUSSIA WITH LOVE

But I'm getting ahead of myself. The point is that even as recently as the 1960s, many distinguished scientists believed there might be technologically advanced alien

---

3 A quick reminder of the electromagnetic spectrum, from long wavelength to short wavelength: radio, TV, microwave, infra-red, visible, ultra-violet, x-ray, gamma-ray.

4 On second thoughts, aliens may be exactly who David Lynch was writing for.

societies within our very own solar system, let alone the galaxy. We have sent surprisingly few radio messages specifically with the intention of making contact with aliens, and the very first, the so-called Mir Message, targeted Venus. Composed in Morse code, and transmitted on 19 November 1962 from a radar dish in Ukraine, it said simply 'MIR, LENIN, SSSR'. In case you are wondering, Mir is Russian for 'peace' and SSSR was the Russian acronym for the Soviet Union. Again, full marks to the Venutian who worked that out.

But I'm here to tell you that as the 1970s wore on, and our knowledge of the solar system increased, this optimism waned. Apollo was cancelled, as was its follow-up, Orion, which aimed to put men on Mars. Instead we switched our attention to unmanned missions, launching a series of robotic probes. By 1972, the Russians had managed to land *Venera* 7 on Venus, and we knew for sure that not only was the surface temperature a face melting 500°C, but its carbon dioxide atmosphere was so thick the air pressure was more than ninety times that of Earth. Three years later, *Venera 9* sent the first black and white photos of the Venusian skyline. It looked like an abandoned slate quarry.

It got worse. NASA's *Mariner 10* made a fly-by of Mercury, the closest planet to the Sun, in 1973. Whereas Venus is practically a twin of the Earth, Mercury is just a little larger than the Moon. As you might expect, it turned out to have no atmosphere, and was pock-marked with craters, indicating that, like the Moon, its core was cold.[5] Less

---

5 Craters tend to indicate that there is no plate tectonics, and therefore no molten core within a planet. One of the shocks of the recent New Horizons fly-by was how few craters Pluto has on parts of its surface. It's too small to have retained much heat from when it was formed, or to have enough radioactive material in its core to drive plate tectonics, and it isn't heated by tidal forces like the moons of the four gas giants.

expected was that, unlike the Moon, it had a weak magnetic field, partially shielding it from the solar wind, but with surface temperatures that regularly shot up to 400°C it was not the kind of place you'd want to call home.[6]

By the time *Viking* landed on Mars, I was ten. That really was a blow. Mercury, Venus and the Moon looked inert, even from Earth, but Mars was different. It was red, the colour of iron, earth and life. When the camera powered up, would a herd of bouffant-haired crabs scuttle for cover? Sadly not. The Red Planet did have a thin atmosphere, so there was a sort of pinkish daylight, but as far as habitability went, that was about it. Mars was a desert.

VOYAGER'S REST

For me, the final nail in the coffin came with the Voyager missions. Throughout the 1980s, wondrously depressing photos emerged as these twin probes flew past Jupiter and Saturn, and *Voyager 2* then continued further, past Uranus and Neptune. Beautiful as each of these four giant balls of gas was, with no solid rock and no liquid water, how could life ever take hold?

We had higher hopes for their rocky moons, but they too were cruelly dashed. The Galilean moons of Jupiter – Io, Europa, Ganymede and Callisto – ranged in size between the Moon and Mercury, and were every bit as barren. There were a couple of surprises. Io had active volcanoes, busy spewing sulphurous gases, and Europa was as smooth as a billiard ball, but that was about it. With no atmosphere, out

---

6 I know; Mercury is closer to the Sun than Venus, but colder. The reason is that it has no atmosphere and therefore no greenhouse effect.

in the freezing boondocks of the solar system they were a biological non-starter.[7]

With the Jovian moons out of the running, attention turned to Saturn. One of the main objectives of the Voyager mission was to investigate Titan, thought at the time to be the largest moon in the solar system.[8] *Voyager 1* plotted a course a mere 6km from its surface, but saw only an impenetrable haze. That meant Titan, unique among moons, had an atmosphere, but it also meant we had no idea what was going on underneath. It seemed pointless for *Voyager 2* to follow up, so instead it was diverted to take a look at Uranus and Neptune. On its way, it managed to grab a photo of another Saturnian moon, Enceladus, which appeared to be a lump of solid water ice.[9]

If the moons of Jupiter and Saturn are chilly, those of Uranus and Neptune are bone-numbing. In January 1986, *Voyager 2* made it to Miranda, a tiny world less than a seventh of the size of our own Moon, with an average surface temperature of −210°C. It had a truly bizarre surface, made up of a patchwork of cratered and smooth sections, leading some to dub it 'Frankenstein's Moon'. Either Miranda had been smashed and hastily reassembled following an impact, or somehow the gravitational pull from Uranus was warming its interior, giving it the icy equivalent of plate tectonics.[10]

7 The four Galilean moons are Jupiter's largest – hence Galileo being able to see them in the first place. Thirteen moons were known at the time of the Voyager mission, which discovered three more. Today the tally is sixty-seven moons.

8 It turned out to be slightly smaller than Ganymede, the moon of Jupiter.

9 Saturn has sixty-two moons, of which seven are big enough to be spherical under their own gravity. They are, in order of increasing orbit, Mimas, Enceladus, Tethys, Dione, Rhea and Iapetus. Titan, the giant, sits furthest out.

10 Uranus has twenty-seven moons, named after characters from Shakespeare, the largest being (usual drill) Puck, Miranda, Ariel, Umbriel, Titania and Oberon.

Finally, in the summer of 1989, *Voyager 2* made it to Triton, by far the largest of Neptune's fourteen moons and three-quarters the size of our own satellite. Like Miranda, Triton was truly alien, and not in a good way. Even its orbit was peculiar. As you may know, the planets and their moons all tend to spin and orbit in the same direction; anti-clockwise if you are looking down on the solar system from above. Not so Triton, which orbits Neptune clockwise, betraying the fact that it isn't a homegrown moon, and was most likely kidnapped from the band of icy rubble outside Neptune's orbit known as the Kuiper Belt. Although small in comparison to Earth, Triton was geologically active with very few craters, and had a surface made of solid nitrogen. Unsurprisingly, it was also one of the coldest places in the solar system, with a temperature of −240°C.

And that was about it. In the 1950s, we had dreamt of battling angry expat Martians and being seduced by blonde-haired Venusians in a tropical paradise. By the end of the 1980s, it was painfully clear that we were going home from the party on our own. Summing it all up was the image *Voyager 1* took on 14 February 1990, as it looked back at the solar system from an orbit halfway across the Kuiper Belt. In a vast expanse of lifeless black, Earth appeared as a single fragile pixel; what Carl Sagan famously called the 'pale blue dot'. His words are so well turned they are worth repeating.

The Earth is the only world known, so far, to harbor life. There is nowhere else, at least in the near future, to which our species could migrate. Visit, yes. Settle, not yet. Like it or not, for the moment, the Earth is where we make our stand. It has been said that astronomy is a humbling and character-building experience. There is perhaps no better demonstration of the folly of human conceits than this distant image of our tiny world. To me, it underscores our responsibility to deal more kindly with one another and to preserve and cherish the pale blue dot, the only home we've ever known.

TRAINING THE *BEAGLE*

The Space Age, which had begun with such optimism, had ended on a cosmic downer. We were alone. Commerce, not exploration, became the driving force behind space science, and the satellite industry boomed. Public interest in space waned, and at one or two dinner parties in north London which I had the misfortune to attend during the early noughties, intelligent and educated people expressed doubt that we had landed on the Moon at all. Internet rumours suggested that mankind's greatest achievement had been a US government hoax, staged in a movie studio by Stanley Kubrick and shot with TV cameras in a bid to demoralise the USSR and to win the Cold War. The astronauts hadn't risked their lives; they had all been fakers.

If I had to pick a low point, for me it would be the launch of the *Beagle*, the life-seeking robot lander that formed part of the European Space Agency's 2003 Mars Express mission. Just as Darwin's voyage on HMS *Beagle* had inspired his theory of evolution, so it was hoped that this plucky little sniffer dog would root out signs of Martian life and rewrite the rules of biology. With a call sign composed by Blur, and a Damien Hirst spot painting as a test card for its on-board video camera, the *Beagle* was basically Britpop on steroids, and about as long lived.

To be fair, it had an ingenious design. Its mother ship, *Mars Express*, had only ever been intended to be an orbiter rather than a lander, but thanks to the charisma and media savvy of the UK's Colin Pillinger, the ESA high-ups were outmanoeuvred and passage secured for a stowaway roughly the size of a bin lid. After jettisoning from the *Mars Express*, two consecutive parachutes would slow the *Beagle*'s descent, and three airbags would cushion its landing. Once on the ground, the airbags would detach, the lid of the main housing would flip open and out would flop four petal-shaped solar panels. A mechanical arm would then emerge, like the stigma from a giant flower, bristling with sampling tools.

Their variety was impressive. As well as the aforementioned video camera there was a microscope, a rock grinder and corer, a wind sensor, a wide-angle mirror and a telescopic drill called the Mole which was capable of digging up to 1.5m into the Martian soil. Once a sample had been collected, the mechanical arm would then manoeuvre it into an inlet port in the central housing, ready for a well-equipped on-board lab to identify exactly what types of molecules were present. If there was – or ever had been – life on Mars, there was a good chance that the *Beagle* would be able to find it.

Touchdown was planned for Christmas Day 2003, and at the allotted hour patriotic Britons waited patiently by their radios and television sets listening out for the otherworldly strains of Blur's call sign. Instead there was silence. The *Beagle* had vanished without trace. The smug mediarati of north London were right, it seemed. The Moon landings were fake, and the ineptitude of the *Beagle* was but so much grist to their infuriatingly self-satisfied mill.

Yet help was at hand. As our interplanetary odyssey languished in the doldrums, a sudden gust of enthusiasm blew in from the most unexpected quarter, driving all before it. While space scientists had focused all their efforts on nearby planets and drawn a series of depressing blanks, their colleagues in the altogether less glamorous world of microbiology had been quietly coming up trumps. Because as it turned out, something akin to aliens had been found, and in the most unlikely of places. They were right here on Earth.

LIFE, BUT NOT AS WE KNOW IT

Tom Brock loved the outdoors. Canoeing and backpacking were a favourite, and in July 1964 he paid a visit to Yellowstone National Park in Wyoming. This, of course, is the home of the famous geyser Old Faithful, which every hour and a half spouts scalding hot water over 150ft in the air. It wasn't this attraction that caught Tom Brock's eye, however; it was the multicoloured scum in the hot springs nearby. Luckily for us, Tom Brock wasn't just an outdoorsman, he was a microbiologist. He knew a microbial mat when he saw one, and he also knew that they shouldn't be growing in near-boiling water.

A microbe is the technical name for a single-celled organism such as a bacterium. As the name suggests, individual microbes are too small to be seen with the naked eye, but, given a suitable environment, they are more than happy to club together to form what is known as a mat. The ones at Yellowstone often contain pigments such as chlorophyll, which you'll know is green, and carotenoids, which can be anything from yellow to red.

Chlorophylls and carotenoids are key players in photosynthesis, being the process whereby microbes, plants and algae use the energy of light to build long-chain carbon molecules from carbon dioxide, known in the trade as carbon fixing.[11] The effect at Yellowstone can be spectacular, particularly at the Grand Prismatic Spring, where the deep blue of the central basin is surrounded by concentric circles of green, yellow, orange and red microbial mats as the water shallows out.

Microbes, of course, are living things, and the conventional wisdom at the time was that they should only exist within a narrow range of temperature. After all, all living things are made of proteins and contain water. Freeze them and they'll go solid. Heat them and their proteins will start to break apart, or denature, a process which in the everyday world we call cooking. Warm your meat or your microbes to anything above 60°C, and you can expect even the most resilient proteins to become gelatin in your hands.

That, at least, was what we believed back in the 1960s. Yet to Tom Brock's astonishment, in the broiling pools of Yellowstone, microbes were positively thriving. Soon he had shifted his research to what became known as 'extremophiles' – organisms which love extremes. There seemed to be no limit to their audacity; during successive visits to Yellowstone throughout the mid- to late sixties, Brock and his research team found strains of bacteria in the Yellowstone pools that thrived at temperatures as high as 90°C.

Extraordinary as this news was, it was bewilderingly slow to catch on. Truth be known, when it comes to reality, we humans are not the most reliable of creatures. Not only are we capable of seeing things that aren't there, but we are also capable of not seeing things that are. There were two

---

11 Fixing is basically the action of taking any gas from the air and converting it to a solid or liquid form. Nitrogen-fixing bacteria, for example, take atmospheric nitrogen gas and convert it to nitrates, which are then hungrily consumed by plants. Soon we shall meet manganese-fixing bacteria, for which I make no apology.

million visitors[12] to Yellowstone Park in 1964, all of them gazing with wonder at the multicoloured hot springs. Many of them must have been scientists, and one or two may even have been of a distinctly microbiological persuasion. Yet no one other than Tom Brock spotted what seems now to be glaringly obvious: the scalding waters were festooned with living creatures that really shouldn't have been there.

Brock wasn't slow to publish his findings, but, even so, it wasn't until the late seventies that news of his discovery started to reach mainstream scientific journals. And at that point our story takes another twist. It's one thing when a diligent microbiologist finds some unusual bacteria photosynthesising in a hot spring in Yellowstone Park; it's quite another when geologists stumble across a whole zoo of unfamiliar creatures two kilometres deep in the Pacific Ocean.

THREE MEN IN A DEEP SUBMERGENCE VEHICLE

We should be grateful to *Alvin*, a three-man deep-sea submersible owned by the US Navy, for two reasons. The first is because in 1966 it recovered an unexploded hydrogen bomb from the bottom of the Mediterranean Sea, after a B-52 bomber collided with a tanker plane while refuelling in mid-air. In that case, it may well have averted nuclear Armageddon and the end of the human race. In the second, some would say it found the very place that the ancestor of all life – the human race included – got its start.

The theory of plate tectonics was put forward by the German geologist Alfred Wegener in 1922, and, as you probably know, proposes that the Earth's crust isn't static,

12 1,929,300 according to the National Park Service.

but is made up of a patchwork of plates, each of which is moving. The joints between plates tend to be where all the geological action happens, in the form of volcanoes, islands, mountains and trenches. Exactly what you get depends on what's on top of the plates – ocean, for example, or a continent – and whether they're being pushed together, pulled apart or are slipping side by side.

At least, that's how it usually works. Sometimes, however, you get volcanoes in the middle of a plate, well away from the edge. In these cases, there seems to be something deep beneath the crust, a 'hot spot' if you like, which the plate is riding over. As the plate moves, the hot spot punches a series of volcanoes up through it. The Hawaiian islands are the classic example, sitting as they do right in the middle of the Pacific plate, which is currently moving north-west towards Eurasia. As fresh Pacific plate moves over the hot spot, plume after plume of hot magma shoots up through it, creating a chain of volcanoes on the ocean floor. The tips of these volcanoes form the Hawaiian islands.[13]

Another example, funnily enough, is Yellowstone Park, though in that case we are presently between eruptions, with the last one having taken place around 640,000 years ago. When a volcano erupts, the encircling area can often collapse, leaving a depression known as a caldera, after the Spanish for 'cooking pot'. It's in the caldera from the last eruption at Yellowstone that we now find Old Faithful and the Prismatic Springs. A third example is the islands of the Galapagos, and that's where the plucky little submarine known as *Alvin* comes in.

---

13 If you look at a map of the northern Pacific, you'll see that the Hawaiian islands form one long diagonal chain. The island of Hawaii is the most recent addition, and has already moved off the hot spot. A new island, the Loihi Seamount, is forming underwater about 35km off Hawaii's coast.

THE GARDEN OF EDEN

On 8 February 1977, *Alvin* set sail from Panama aboard a purpose-built catamaran named *Lulu*, heading for a deep-ocean volcanic ridge just north of the Galapagos Islands known as the Galapagos Rift. Once there, the plan was to try and find hot springs. The saltiness of the world's oceans suggested they were getting a supply of salt water from somewhere, and the smart money said that some-where down on the sea floor there had to be the equivalent of Yellowstone's pools and geysers, pumping out salts and minerals. Nevertheless, at the time of *Alvin*'s dive, no one had yet found a real hydrothermal vent that would settle the issue one way or another.

The previous summer, a survey of the Galapagos Rift had used an unmanned deep-sea camera to hunt for hot springs, but without success. At one point, however, the umpteen photos of barren sea floor were interspersed with a few brief shots of a pile of dead white clam shells, along with a beer can. The team assumed it was just rubbish thrown overboard by a ship having a party, and named the site 'Clambake'. After all, nothing could possibly be living at that depth, because there was no light.[14] Without light, there would be no plants, algae or bacteria. And without them there was nothing for anything else to eat.

How wrong they were. On 17 February 1977, *Alvin* took a dive, piloted by one Jack Donnelly and carrying two geol-ogists, Jack Corliss and Tjeerd van Andel. As they neared the ocean floor, the water began to shimmer. Sure enough, hot

---

14 As you dive, the pressure increases by roughly 1 bar every 10m. Atmospheric pressure is 1 bar.

water was pumping out of the dark volcanic rock, and form-ing black sulphurous clouds as it cooled, earning such vents the nickname of 'black smokers'. But that was far from all. As *Alvin*'s searchlights scoured the surrounding rocks, they revealed a ghostly menagerie of extraordinary creatures. There were giant white clams, white crabs and even a purple octopus, all very much alive. Confused, Corliss picked up the acoustic telephone and called his graduate student Debra Stakes, above them on board *Lulu*. 'Isn't the deep ocean supposed to be like a desert?' Corliss asked. When Stakes confirmed that it was, a puzzled Corliss replied: 'Well, there's all these animals down here.'

It got weirder. Subsequent dives revealed more hot springs, and even more strange creatures. At another spring they found an orange animal that resembled a dandelion; at another that they breathlessly christened the Garden of Eden, they found a forest of giant tubeworms with bright red tops, swaying in the water like a field of flowers. They did their best to collect specimens, but being a geology expedition they had little in the way of formaldehyde to preserve them in. Instead, they used the next best thing: some bottles of Russian vodka they had bought in Panama. Tjeerd van Andel began to lose interest in his original goal of finding hot springs, and lay awake at night, his mind buzzing with questions. Where had these creatures come from? What could they possibly be eating?

Two years later, in 1979, a team of biologists returned to find out. *Alvin* was modified with a new collecting basket and a second mechanical arm, and fitted with a movie camera. On each dive, the team returned with a zoo of creatures that had never been seen before: new species of mussels, anemones, whelks, limpets, featherduster worms, snails, lobsters, brittle stars and blind white crabs. The delicate orange dandelion-like creature seen on the 1977 dives turned out to be a relative of the Portuguese man-of-war, though it quickly disintegrated after being brought to the surface. Finally, the mystery of what all these creatures were eating was solved by a biologist called Holger Jannasch. At the base of this baroque food chain was a microbe. Rather than getting its energy from sunlight, this bacterium was feeding on a chemical in the vent fluid; specifically, hydrogen sulphide.

Suddenly, all bets were off. If life didn't need light, or moderate temperature, where else on Earth might it be lurking? Suddenly, extremophiles seemed to pop up everywhere we looked. We found microbes in nuclear reactors lapping up radiation ten times stronger than that which would kill the hardiest cockroach. We found both fish and microbes thriving under extraordinarily high pressure 11,000m underwater in the Challenger Deep and the Marianas Trench. We found microbes and fungi bathed in acid so strong it has a pH of zero. We even found bacteria that live inside rocks.

All of which raised an interesting question: who was really living at the extremes, them or us? To a bacterium which lives in the sweltering heat of a Yellowstone spring, aren't we the extremophiles, able to endure dessicatingly dry surroundings so cold that the little water that is available regularly freezes solid? What was the natural environment of the first life? Had the first cells incubated in the shimmering heat of a black smoker, and only later migrated to cooler, sunlit shallows? If microbes could live in rocks, could they travel between planets on meteorites? Had life begun elsewhere – on Mars, maybe – and taken a joyride to Earth on a space rock?

Life, in short, simply wasn't what we thought it was.

It wasn't delicate, or precious, or in any way predictable. Far from it: it was tenacious, commonplace and infinitely adaptable. Here on Earth, all it seemed to require was water, carbon and a source of energy. Maybe Mars, Venus and Mercury weren't quite the inhospitable deserts we had once feared them to be. If they contained even trace amounts of moisture, they might be home to some sort of bacteria. Could some of those icy moons, orbiting giant gas planets in the outer reaches of the solar system, be habitable after all?

THE ORIGIN OF THE SPECIES

There's something wonderfully poetic about *Alvin* finding a whole new raft of life near the Galapagos, of course, because it was on these volcanic islands that the great Charles Darwin collected the specimens that were to inspire his theory of evolution. The story is worth retelling, not only because Darwin was the astronaut of his day, boldly going where no naturalist had gone before, but also because evolution is such a linchpin in our search for intelligent extraterrestrial life. What follows may seem like a diversion, but by taking it we will have an easier approach to the summit, so here goes . . .

Galápago is a Spanish word meaning 'tortoise', and, according to Darwin's journal, on 18 September 1835 the crew of the *Beagle* brought fifteen giant tortoises on board from Chatham Island,[15] ready to supply a feast. With few natural predators, the animals of the Galapagos were curiously trusting; in fact, hunting and collecting were pretty much the same thing. Darwin himself reports knocking a

15 Now known as San Cristobal.

24

hawk off a branch with the tip of his rifle, and rather surre-
ally recalls midshipman King killing a bird with a hat.

Understandably, these giant tortoises made a strong
impact on our young hero, and he was intrigued by the
observation of the vice-governor of the Galapagos, Nicholas
Lawson, that 'he could, with certainty tell from which island
any one was brought'.[16] In other words, each island had its
own species of tortoise. Sadly, Darwin wasn't that successful
in finding specimens that proved the point. He collected
three giant tortoise shells, each from a different island, but
they were from young animals and there was little to tell
them apart.

Back in London, Darwin presented all his specimens to
the Geological Society of London. The birds he gave to
John Gould of the Royal Zoological Society for examination. Among them were those from the Galapagos, which
Darwin had identified as blackbirds, wrens and finches.
When Gould returned the surprise result that they were
all finches, 'so peculiar as to form an entirely new group,
containing twelve species', Darwin began to formulate an
audacious idea. What if there had originally been no finches
on the Galapagos, which were, after all, relatively new volcanic islands. Could it be that a mating pair of finches had
flown there from the South American coast, and somehow
their descendants on each of the various islands had meta-
morphosed into new species?

To prove the point, Darwin needed to be able to show
that, as with the vice-governor's giant tortoises, each island
was home to a different species of finch. Unusually for the
meticulous Darwin, he had failed to label his own birds
accurately, but fortunately his servant Syms Covington had

16 From *The Voyage of the Beagle.*

not been so sloppy. By combining Covington's specimens with those of the *Beagle*'s captain, Robert Fitzroy, Darwin was able to reconstruct the locations where he had found his own finches. It was true: each island had begat its own species. Gould had returned his result on 10 January 1837. That March, Darwin wrote in a notebook the words that would change the course of biology forever: 'one species does change into another'.

Species could change, but how? For Darwin, the argument went something like this. The mating pair had prospered, and their offspring had populated the various islands. Since reproduction is never exact, within each island population there were a variety of traits. Some finches, for example, had thick beaks, while others had thin beaks. If the seeds on a given island were hard to crack, finches with thick beaks would have a survival advantage, and would therefore have more offspring. Eventually, given sufficient generations, the entire population of finches on that particular island would have thick beaks. Nature, in other words, did not favour all creatures the same. Some she selected, and some she did not. As Darwin put it, species evolved through a process of natural selection.

In July that same year, barely eight months after he had returned on the *Beagle*, Darwin picked up his notebook and wrote the words 'I think', and below them sketched the first tree of life. Starting from a single trunk – the first living creature – he drew branch after bifurcating branch, with each new outgrowth representing a new species. It was a simple drawing, but its implications were profound. Starting with a single organism, life on Earth had evolved into an ever-increasing number of species. Take any two living things, the figure said, and you could trace back their lineage to find a common ancestor. All life on Earth was one.

IT'S ALL IN THE GENES

That's such a piquant thought it's worth taking a moment to digest it. Every single living thing on the planet is related to every other living thing. Not only are you a descendant of your great-aunt Ada, but you are a distant cousin of a flatfish, and a kinsman of an amoeba. The creatures that we find at black smokers, strange as they are, perch on the same tree of life that we do, as do the most bizarre fossils we have ever found, those of the Ediacarans.[17]

Completely central to Darwin's theory of evolution by natural selection, of course, is the concept of inheritance: the passing of traits from parent to offspring. In Darwin's day, the mechanism of reproduction was unknown; today we understand that every organism on Earth carries its own blueprint in every cell of its body, coded into the long-chain carbon molecule known as deoxyribonucleic acid, known as DNA. In short, you resemble your parents because you inherited their DNA.

Or to be more accurate, you inherited almost all of it. The system by which DNA is copied isn't perfect, and that's vital; it's this less than perfect copying that gives rise to what Darwin called 'variation': the appearance of a new trait in the offspring that wasn't inherited from its parents. Most of the time the new trait makes no difference. Sometimes it is harmful, and the offspring will be less likely to reproduce as a result, meaning the new trait dies out. In some rare cases, however, it bestows a survival advantage, and increases the chance that the offspring will, as Daft Punk might put it, 'get lucky'.

17 The Ediacarans are the first known complex life forms, ruling the planet some 575 million years ago, and lacked eyes, mouths and limbs. More on them later.

Traits are coded by small sections of DNA known as 'genes'.[18] Take any trait – the thickness of a finch's beak, say – and it is possible to identify the genes which control it. Indeed, as the British biologist W. D. Hamilton showed, it is at the level of genes that the struggle for survival is best understood, rather than that of an organism or species. Put simply, genes are doing it for themselves.[19] All that matters to your genes is that they make as many copies of themselves as possible; we host organisms are simply a means to an end.

I'M A MAC, ZARG IS A PC

So there we have it. Evolution is the key that unlocks the mystery of life-as-we-know-it, and is able to explain two extraordinary and seemingly unrelated facts. Firstly, the older the fossils we dig up the more primitive the life forms we find. No one has yet found a mastodon in the same layer of rock as a trilobite, nor do we ever expect them to. Speciation – the process by which natural selection creates two species where previously there was merely one – is irreversible, and the total number of species,[20] both living and extinct, can only increase with time.

And, secondly, it explains the extraordinary similarity between the different branches of life-as-we-know-it. To use a computer analogy, everything we find is a Mac; nothing is a PC. Every creature on Earth is made up of one or

18 Just so as you know, the word 'gene' is sometimes used in a different sense, to mean 'bit of DNA that codes for a given protein'.

19 Hamilton's work has been rather brilliantly popularised by Richard Dawkins, who summed the whole thing up with one pithy phrase: 'the selfish gene'.

20 Estimated at 8.7 million by the United Nations Environment Programme in 2011, give or take around 1.3 million.

more cells, relies on water as a solvent, stores its blueprint in DNA and burns carbohydrate to release energy. Every molecule of living protein is made from the same twenty amino acids,[21] and every molecule of living DNA is coded using the same four nucleobases.[22] On an even deeper level, you might say that all life-as-we-know-it is carbon-based, because almost every single molecule that you can think of that has a biological or biochemical function is a compound of carbon.

So what does this mean for our search for intelligent alien life? Well, as the Galapagos giveth, so the Galapagos taketh away. On the one hand, its black smokers show us that there's no limit to the kind of environments where life might thrive. On the other, its finches tell us there's only one kind of life on Earth. So are we alone or not? What if life-as-we-know-it is a colossal fluke, a one in a gazillion random event, never to be repeated? One thing is for sure: if we had just one other example of a second tree of life here on Earth, we'd be a lot more confident that life is common in the galaxy. And yet there's nothing. Or is there? It's time to talk about desert varnish.

LIFE IN THE SHADOWS

The godfather of desert varnish research was the pioneering German naturalist and explorer Alexander von Humboldt. In 1799, during his groundbreaking expedition to South

21 As the name suggests, amino acids are made up of an amine (NH_2) group attached to an acid.

22 The common nucleobases are made up of one or two rings of carbon atoms, with a couple of nitrogen atoms inserted into the ring. The ones we find in DNA are guanine, adenine, thymine and cytosine.

America, he noticed a strange metallic coating on the granite boulders in the rapids near the mouth of the Orinoco River in north-eastern Venezuela that made them appear 'smooth, black, and as if coated with plumbago'.[23] Intrigued, he had the coating analysed by one of the leading chemists of the day, Jöns Jacob Berzelius,[24] who informed him that it was made up of manganese and iron oxides. That was puzzling, because granite contains only small amounts of manganese and iron.[25] Where were these metals coming from? Presumably from the waters of the Orinoco, but what was attaching them to the rocks?

In *The Voyage of the Beagle*, Charles Darwin also describes an encounter with mysterious rock coatings. His were found below the tide line on a beach on the coast of Brazil and were a 'rich brown' in colour. Darwin had a bit of a thing about Humboldt, and one of his prized possessions on the *Beagle* was a seven-volume translation of his hero's *Personal Narrative* of his South American voyage.[26] He knew that Humboldt's coatings had been much darker, and wondered if the redder colour of those formed on the beach in Brazil was because they contained less manganese and more iron. Struck by how they 'glitter[ed] in the sun's rays', he too was puzzled by what could be causing them, remarking that 'the origin ... of these coatings of metallic oxides, which seem as if cemented to the rocks, is not understood'.

Both Humboldt and Darwin found their coatings in the tropical climate of South America, but – perhaps

23 'Plumbago' is Humboldt for 'graphite'.

24 Most famous for coming up with the letter symbols for the chemical elements.

25 Most granite contains iron oxide at 1.68 per cent by weight, and manganese oxide at 0.05 per cent by weight.

26 Full title: *Personal Narrative of Travels to the Equinoctial Regions of America During the Years 1799–1804.*

unsurprisingly given the name – it turns out that what we now call desert varnish is found just as often in arid surroundings. The sandstone deserts of the Colorado Plateau in the south-western United States are a classic example, where the shiny black patina on the rocks has often provided a handy surface into which Native Americans could scratch their art. Within the canyons, the swathes of varnish can be particularly striking, sometimes covering entire walls, or forming vertical stripes alternating between black and red and tan.

We've learned a bit more about desert varnish since the time of Humboldt, but not much. As Darwin suspected, its colour does indeed vary from red to black depending on the relative quantities of iron and manganese oxide, with varnishes with equal amounts of both appearing tan in colour. We know that the varnish also contains silica in the form of clay, and that it grows more readily on rocks and walls that are intermittently wet and get good sun, as if the rapid drying of water somehow helps the varnish grow. We know that the varnish accumulates slowly, growing by less than the thickness of a human hair every thousand years. And we also know – and this is where it gets controversial – that it contains microbes.

Barry DiGregorio, for example, an Honorary Research Fellow at the Buckingham Centre for Astrobiology in the UK, thinks that these microbes are photosynthesising manganese-fixing bacteria, and that the varnish is, like the blooms that Brock found in the hot springs of Yellowstone Park, a type of microbial mat.[27] On the other hand, Randall Perry, a researcher in the Earth Science Department of

---

27 It is thought that manganese-fixing bacteria were a crucial stopping-off point in the evolution of photosynthesis; indeed, all plants today use manganese as a building block of chlorophyll.

Imperial College London, believes that it's the clays in the varnish that are doing all the heavy lifting. He thinks they react with moisture to form a gel, which then traps all sorts of other stuff including stray microbes, as well as catalysing some pretty funky chemistry which concentrates the metal oxides. When this gel dries in the sun, it hardens into a varnish.[28]

Who's right? Well, maybe neither, says Carol Cleland, a philosopher at the University of Colorado Boulder in the US, and an affiliate of the NASA Institute of Astrobiology. She is fascinated by the fact that we don't know whether desert varnish is chemical or biological. As she points out, it's a stretch to see how rocks could become coated with metal simply as a result of chemistry, but on the other hand, when we excavate desert varnish – if you can call scratching away at something a hundredth of a millimetre thick an excavation – we don't find lots of cells, just a few fragments. What, she asks, if desert varnish is another kind of life entirely?

At first glance, that might seem slightly unhinged, but she has a point. After all, extremophiles have been around for billions of years, but it took Brock to notice them; once he had, we started to find them everywhere. In 2005, Cleland proposed the existence of a 'shadow biosphere'; a microbial ecosystem living in parallel with our own, but that we have yet to identify. As she rightly points out, all of our tests assume that there is only one kind of life: our own. What if there's something out there that doesn't have the same DNA code, or uses different amino acids to build its proteins? What if it doesn't have DNA or proteins at all? Or cells? Or

---

28 The gel is silicic acid, and the reaction can be written as clay + water → silicic acid, or if you are that way inclined, $SiO_2 + 2 H_2O → H_4SiO_4$.

is based on something other than carbon? How would we recognise it?

Cleland feels that we should be actively searching for life-as-we-don't-know-it, and that desert varnish is a good place to start. Another is manganese nodules, the strange metallic boulders that populate the seabed in many of our oceans, and which always remind me of the egg hatchery in the movie *Alien*. These, again, are assumed to be the result of chemistry rather than biology, but how can we be sure? Once we can rid ourselves of our preconceptions of what life should look like, maybe we will start to find it everywhere; even in the hot springs of Yellowstone National Park.

Of course, all of this begs a question: if we do find a second genesis of life here on Earth, what do we call it? Seeing as it's from Earth, the word 'alien' doesn't seem right. The cosmologist Paul Davies has suggested ditching the word 'alien' altogether, and using the term 'weird life' to describe anything that doesn't share a common origin with life-as-we-know-it; he has even proposed a 'mission to Earth' to seek it out. And what do we call the kinds of life that are emerging from the ever growing field of synthetic biology? Is that weird too, or just weird-ish?

IN OUR BACKYARD

However you slice it, one thing is certain: the discovery of extremophiles has opened our minds to the idea that our own solar system might be teeming with life after all. There may not be herds of wildebeest sweeping majestically across the plains of Jupiter's moon Europa, but there may well be blooms of micro-organisms in the giant oceans we now know to be hidden beneath its icy crust. Likewise

Ganymede, the largest of Jupiter's Galilean moons, is now known to be hiding a salt-water ocean sandwiched between layers of ice. Knowing what we now know about life's appetite for weird and wonderful environments, what might be lurking in its depths?

Remember Enceladus, the tiny moon of Saturn which appeared to *Voyager* to be a solid lump of ice? Well, in July 2005, NASA's *Cassini* spacecraft arrived to finish the job. To everyone's surprise, Enceladus wasn't a cold, dead world after all, but violently active. At its south pole was a giant volcanic hot spot, from which plumes of ice particles and water vapour were erupting hundreds of miles into space; in fact, it's this firehose of material that supplies Saturn's vast outer ring. By 2014, further measurements by *Cassini* had confirmed what many suspected: beneath the pack ice at Enceladus' south pole was a giant superheated ocean.[29]

Although Mars appeared lifeless to the *Viking* lander in the late seventies, circumstantial evidence has grown that it too was home to microbial life in the past, and may still be today. Thanks to recent missions like the Mars Reconnaissance Orbiter (MRO) we know that the Red Planet had copious amounts of water until as recently as two billion years ago. That means conditions on Mars were right for life-as-we-know-it round about the same time as they were on Earth; some have even suggested that our kind of life seeded on Mars first, and then came to Earth on a meteorite. As recently as 2015 NASA's *Curiosity* rover found nitrates, a compound essential to many forms of life, and the MRO even found evidence of running water.

Last, but most definitely not least, is the glorious Titan.

---

29 By the way, in 2013 a water plume was also reported on Europa. It now seems as if that was caused by some sort of freak event, such as a meteorite impact.

Again, it may not have been love at first sight, but he's growing on us. *Voyager 1*, as you will remember, saw nothing but an orange methane smog, and *Voyager 2* decided it had better things to do than follow up. Yet many wouldn't let it go. Just as the temperature of the Earth sits near the triple point of water – that is, the temperature at which water at atmospheric pressure can exist as a solid, a liquid and a gas – *Voyager 1* had confirmed what we had suspected since the middle of the twentieth century: that Titan was at the triple point of methane, with a methane-rich atmosphere.[30] When NASA's *Cassini* spacecraft made its flight to Saturn, hitching a ride was a probe named *Huygens*, built and paid for by the European Space Agency.[31]

If it's weird life we're looking for, the images beamed back by *Huygens* showed us that Titan is just the place to find it. Martian meteorites reach us so frequently, and Mars' early climate was so similar to ours, that any life we find there is quite likely to share a common genesis, or at least a similar biochemistry. And that's leaving aside the issue that the *Viking* lander wasn't sterilised properly, and may have contaminated Martian soil with earthly microbes. But Titan is a whole other deal. This is a really big moon with a proper atmosphere, with methane clouds, methane lakes, methane ice, and maybe even methane snow. What kind of abominable methane snowman might be lurking there?

Abominable microbial methane snowman, I should say. While our solar system no longer appears to be the graveyard we once thought, no one is expecting to find anything within it other than simple, single-celled life.

30 For the physicists, Titan's surface temperature is 94K, and the partial pressure of its methane is 117 millibars; the triple point of methane at that pressure is 90.7K.

31 The Dutch astronomer Christiaan Huygens discovered Titan in 1655.

That's extraordinarily exciting just in itself – imagine what we could learn from just one weird microbe – but its implications would be even more exciting still. It would mean that biology is as universal as chemistry. And where there's biology, there's evolution. And where there's evolution, there's complex, intelligent life.

CHAPTER TWO

SETI

In which our author introduces us to the scientific search for alien radio signals, known as SETI, acquaints us with the famous Drake Equation and investigates the strange phenomenon of UFOs.

It was the summer of '67, and Jocelyn Bell Burnell's[1] wildest dreams were about to come true. As a research student under the eminent astronomer Antony Hewish, she had spent the previous two years helping build a brand new radio telescope at the Mullard Radio Astronomy Observatory just outside Cambridge. Now that same machine was ready to explore the virgin heavens, and she alone would be the confidante of its secrets.

A radio telescope, of course, makes an image of distant stars and galaxies using the radio waves they give off. Not that the average radio telescope looks particularly like the sort of thing Admiral Nelson defiantly held up to his blind

1 Then Jocelyn Bell.

eye; in fact, the one that Bell Burnell was in sole charge of looked more like two rugby pitches laid end to end and covered in TV aerials.

This being the mid-sixties, the output of Bell Burnell's telescope was not the hard drive of some freon-cooled supercomputer but four three-pen chart recorders, whose inkwells and chart paper needed replenishing every morning and which produced 96ft of recorded chart paper every day. And after a few weeks spent getting her eye in, Bell Burnell noticed something very strange indeed.

The telescope she was using had been purposely designed to investigate a newly discovered kind of radio source called a quasar. A quasar is a galaxy at the very beginning of its life, blasting out radio waves as the supermassive black hole at its centre feasts on extremely high temperature gas and dust. Bell Burnell was soon able to pick out good candidates for quasars, and to discard unhelpful noise from earthbound radio sources such as dodgy spark plugs on passing mopeds on the nearby A603. But there was another type of signal she could not account for: a rapid juddering of the chart pens that produced a quarter-inch or so of what she called 'scruff', that cropped up roughly once every 3000ft of chart paper.

It didn't take Bell Burnell long to work out that the 'scruff' must be coming from the same patch of sky; in fact, it was in step with the distant stars, implying it was well outside the solar system. To see the 'scruff' in more detail, Bell Burnell began to set the chart paper to run faster every time the telescope scanned that particular corner of the cosmos. The results were extraordinary. The 'scruff' resolved into a signal. There on the chart paper was a regular pulse, with each pulse precisely 1⅓ seconds apart.

Bell Burnell was stumped. What on earth, or rather what not-on-Earth could it be? Stars and galaxies glow, they don't pulse. Pulses mean life. And then an extraordinary thought struck her. Could this be a message from an alien civilisation?

FLYING SAUCERS

We'll return to the story of Jocelyn Bell Burnell and her mysterious radio pulses at the end of the chapter; suffice it to say that her exemplary detective work and scientific nous will provide an exquisite counterpoint to some of the undeniably entertaining but rather bonkers UFO stories which follow.

I can imagine that some of you already feel offended. You picked this book up in the hope that it would be about UFOs, and you feel like you've been short-changed. Maybe you've seen a UFO – which simply stands for Unidentified Flying Object – or you know someone else who has. I think it's worth getting something clear from the start; I have an open mind. To me, that's what science is all about. It also means not accepting a theory as correct unless it is supported by high-quality evidence, no matter how much you want it to be true. UFO stories are great fun, and I enjoy them as much as anyone; I just don't believe that they have much to do with real-life extraterrestrials.

That said, I think it's worth giving a potted history of the UFO phenomenon so that we can put the real science of alien life in context. So many people see UFOs, report contact from UFOs and recount abduction by UFOs that something must be going on. What is that something and when did it start?

The term UFO was coined by the US Air Force in the 1940s to describe anything seen in the sky that cannot easily be explained in terms of known craft or natural phenomena. Towards the end of the nineteenth century, witnesses began to report sightings of alien airships; these were then followed by sightings of alien rockets in the first part of the twentieth century. But the phenomenon as we know it really took off with the appearance of that alien design classic, the flying saucer.

So when did the first flying saucers appear? Strangely enough, there's a precise answer to this question: Tuesday, 24 June 1947. Because that was the year amateur pilot Kenneth Arnold made an extremely memorable business trip.

UP UP AND AWAY

At two o'clock that afternoon, the thirty-two-year-old Arnold took off from Chehalis in Washington State in a three-seater, single-engine Callair, heading for Yakima some 120 miles due east, on a course that would take him past Mount Rainier in the Cascade Mountains. A military plane had crashed on the mountain the previous winter, killing thirty-two marines, but because of the snow the wreck had never been found. It was a beautiful clear day, and with the snow receding – and the incentive of a five thousand dollar reward – Arnold decided to go check it out.

As his plane emerged from searching one of the canyons at the foot of Mount Rainier, Arnold saw a bright blue flash, and thought for a moment that he must have caught a reflection from a plane very close by. Alarmed that he might be on a collision course, he scanned the sky around him, but could see no craft in the immediate vicinity. A second bright blue flash then lit up his cockpit with the brilliance of a 'welder's arclight', and in the distance he saw 'to the left of me a chain of objects which looked to me like the tail of a Chinese kite, kind of weaving and going at terrific speed across the face of Mt. Rainier'.

At the centre of each craft there was a bright blue light, pulsing in a way that he later said reminded him of a human heartbeat. From their diagonal formation and high speed he thought at first they must be some kind of military planes. For Arnold, the really strange thing was that the craft didn't have any tails; silver in colour, they looked 'something like a pie plate that was cut in half with a sort of convex triangle in the rear'. Never having seen anything similar before, Arnold decided that the tails must be concealed with camouflage paint, and 'didn't think too much of it'.

Being a good pilot, Arnold glanced down at the second hand on his watch and timed the fleet of ships as they made the journey from Mount Rainier to Mount Adams, clocking the trip at one minute and forty-two seconds. That got his attention. The two peaks were separated by some fifty miles, so the speed of the fleet must have been something like twenty-five miles per minute, or 1,500 mph. That was really going some. At the time the air speed record was less than half of that, at around 620 mph. He counted nine craft in total, judging their closest approach to be twenty-three miles, and their wingspan to be at least 100ft across.

After landing at Yakima, Arnold went to the office of his friend Al Baxter, the general manager of a crop dusting company called Central Aircraft. Bemused by the story, Baxter called in two of his flight pilots and a helicopter pilot for a second opinion. The best explanation they could offer was that Arnold had spotted some test missiles from the nearby army air base at Moses Lake.

One of the local papers, however, saw something else in the story. Were the strange craft really military planes, or something else altogether? The day following Arnold's encounter, a very brief column appeared in the *East Oregonian* inaccurately stating that Arnold had observed a 'saucer-like aircraft'. The paper's editor, Bill Bequette, decided to put the story on the Associated Press wire to see if the US military would respond and clear the matter up. The wire stated that Arnold had seen 'nine bright saucer-like objects flying at "incredible" speed . . . ' It seems impossible,' Arnold said, 'but there it is.'

When Bequette got back to the office after lunch, the phone was ringing off the hook.

SURFING THE WAVE

Arnold's sighting caused something of a media frenzy, and launched what is known as 'The 1947 Wave' of UFO sightings. Many of these sightings were also of flying saucers, metallic, disc-like craft that travelled at great speed. Others were of rockets, balloon-like craft and balls of light. Three days after the Arnold incident, a rancher named William Brazel found some strange looking debris on his ranch outside Roswell, New Mexico, and called the local Army Air Force base saying he had found a flying saucer. The base's PR officer passed the report on to the press, and news spread that the US government had recovered a crash-landed flying saucer. The phenomenon spread, and by the end of July 1947 there had been a total of forty-five

UFO sightings across America, seventeen of which were of flying saucers.[2]

After the flying saucer sightings of the 1940s, the 1950s saw the UFO phenomenon ramp up a notch with the emergence of the so-called 'contactees'. These were people who claimed to have communicated with aliens. One notable case was that of George Adamski, who said that an alien spaceship made of translucent metal had landed next to him in the Colorado Desert and a blond Venusian named Orthon had warned him of the dangers of nuclear war via telepathy.

The 1960s then saw the first abduction stories, beginning with Betty and Barney Hill's encounter with a UFO on a night-time drive through White Mountain National Park, New Hampshire, in 1961. Both of them remained troubled by the event, and two years later they recalled under hypnosis that they had been kidnapped by little grey aliens with large black eyes who had, among other things, examined their genitals and showed a keen interest in Barney's dentures.

The crop circle phenomenon then briefly took over in the 1970s, reaching a peak in the late 1980s. In 1991 two Englishmen, Doug Bower and Dave Chorley, admitted responsibility for the hoax. They had been inspired by the 1966 case of the Tully 'saucer nest', when a farmer from Tully, Queensland, Australia, reported seeing a saucer-like

---

2 I have to say, checking data in the world of UFOs has to be one of the most mind-bendingly fruitless activities that has ever befallen me. Wikipedian beware is all I can say. One report I read on the internet quoted 850 sightings in July 1945 alone, which I suspect is closer to the number of media articles than to the number of actual eyewitness accounts, if it is based on anything at all and not simply plucked out of thin air. The figures I've quoted here are taken from the US Air Force's investigation Project Sign, which is available – along with declassified pages from the later Project Grudge and even later Project Blue Book – at http://www.bluebookarchive.org

craft rise up from a swamp and fly away, leaving behind a flattened circular area of grass.

And that, in a nutshell, is the UFO phenomenon. Kenneth Arnold is reported as seeing flying saucers in 1947, followed by a wave of sightings across the US. The first contactees of extraterrestrials appear in the 1950s, followed by the first abductees in the 1960s. Crop circles come and go in the 1970s and 1980s and are shown to be a hoax. UFO reports still occur today, and are now mainly of the abductee type. Clearly something is going on here, but what exactly is it? Could it really be true that aliens are visiting our planet?

THE ALIENS ARE COMING

The science on this is pretty clear: yes, it could. It's true, the distances between the stars in the galaxy are prohibitively large, making any journey between alien worlds a considerable challenge. Our nearest star is a red dwarf called Proxima Centauri, which is 4.24 light years away, and which may or may not have planets, we're not sure. As we learned at the beginning of the last chapter, the furthest mankind has managed to send a spacecraft is the *Voyager 1* probe, which thirty-five years after its launch is only just reaching the edge of our solar system at a measly distance of 0.002 light years.[3] But who's to say what superior technology a long-lived intelligent civilisation might create? Why shouldn't a souped-up alien spaceship be able to fly at an appreciable fraction of the speed of light? Or shortcut

3 In case you're interested, *Voyager 1* is heading in the direction of the snappily named star AC +79 3888 in the northern constellation of Camelopardalis, and is expected to arrive in around 40,000 years' time. That's obviously a lot of time to spend in an economy class seat.

between distant regions of space by harnessing the energy of a star to create an Einstein–Rosen bridge, better known as a wormhole?

OK, the wormhole thing is pushing it. But basic calculations show that even aliens with reasonably fast spaceships could rapidly colonise the galaxy if they so choose. In fact the astronomer Paul Davies has calculated that a single alien civilisation with ships travelling at only (only!) a tenth of the speed of light could cross the entire Milky Way Galaxy within four million years.[4] Unless there's something extremely special about us or the Earth, the Milky Way should be awash with civilisations older and more technologically advanced than ours. Surely one of them would have visited us by now?

THE FERMI PARADOX

The most famous statement of this line of reasoning was made by the eminent Italian physicist Enrico Fermi and is known as the Fermi Paradox. Fermi is something of a hero figure among physicists, a brilliant theoretician who was also a gifted experimental scientist. Awarded the Nobel Prize in 1938, Fermi used the award ceremony in Stockholm as an opportunity to escape Mussolini's Fascist Italy and emigrate to the US with his Jewish wife Laura.[5]

---

4 His calculation assumes that each colony takes 1,000 years to establish itself and that suitable planets are on average ten light years apart.

5 And Mussolini must have been very cheesed off that he did, because it's not too much of an exaggeration to say that Enrico Fermi helped alter the course of the Second World War. As a linchpin of the Manhattan Project, together with Robert Oppenheimer and Edward Teller, Fermi helped develop the atomic bombs 'Little Boy' and 'Fat Man' that were dropped on Hiroshima and Nagasaki, effectively ending the war.

In fact, among scientists Fermi is as famous for his powers of estimation as he is for his contributions to the Manhattan Project and his Nobel Prize. Should you be wondering, the relevance to the search for extraterrestrials is as follows. In 1950, Fermi paid a visit to Los Alamos, where Teller was working on the successor to the atomic bomb, known as the hydrogen bomb. That summer had seen a strange phenomenon in New York: the mass disappearance of public trash cans. It had been a good summer for UFO sightings, too, and as they walked to lunch one of Fermi's colleagues told him of a cartoon that he had seen in the *New Yorker*, where a flying saucer was pictured unloading New York trash cans on its home planet.

Fermi joked that the cartoon presented a reasonable scientific hypothesis, because it explained two separate phenomena. Teller recalls that a serious discussion then followed about whether flying saucers were real, with no one in the group feeling particularly convinced. Fermi then enquired as to the probability of faster-than-light travel. What were the chances that they would see material evidence of a solid body travelling faster than the speed of light within the next decade? Teller put the odds at a million to one; Fermi was more optimistic, putting them at one in ten.

The group entered the Fuller Lodge canteen and settled down to lunch, making small talk as only physicists working on the most powerful weapon in the history of humanity can. Then, halfway through the meal, out of the blue, Fermi suddenly asked, 'Where is everybody?'

His companions immediately grasped that Fermi was talking about extraterrestrials, and burst out laughing. One of the scientists present, Herbert York, recalls that Fermi followed up with a series of calculations on the probability of earthlike planets, the probability of life given an earth, the probability of humans given life, the likely rise and duration of high technology, and so on. He concluded on the basis of such calculations that we ought to have been visited long ago and many times over. As I recall, he went on to conclude that the reason we hadn't been visited might be that interstellar flight is impossible, or, if it is possible, always judged to be not worth the effort, or technological civilization doesn't last long enough for it to happen.

In other words, after running the numbers Fermi concluded that intelligent civilisations must exist, but for one reason or another they are staying put. But if they aren't going to come calling on us, how do we get to meet them? The answer: radio waves.

OPENING THE WINDOW

At least that was the conclusion of a paper entitled 'Searching for Interstellar Communications' by Giuseppe Cocconi and Philip Morrison of Cornell University, published on 19 September 1959 in the scientific journal *Nature*.[6] They pointed out that there happens to be very little noise on Earth over a certain range of the radio spectrum, known in the trade as the microwave window. For frequencies below this window there's lots of noise because of absorption and

6 Giuseppe Cocconi would later become one of the leading lights of CERN, while Philip Morrison famously narrated the outstanding science short *Powers of Ten*. If you've never seen it, you have a rare treat in store at http://www.powersof10.com/film. Enjoy!

emission by interstellar gas and dust, and above it there's lots of noise because of absorption and emission by the Earth's atmosphere. If aliens wanted to contact us, they reasoned, they would most likely send radio waves tuned to sit in this particular window.

THE DRAKE EQUATION

Frank Drake was nervous. It was November 1961, and in a few days' time he would be hosting the first conference of SETI, the Search for Extra-Terrestrial Intelligence. Among the attendees would be one of his heroes, the Russian émigré astrophysicist Otto Struve. During his navy scholarship in electronics at Cornell in the early fifties, Drake had attended one of Struve's lectures, and had been gripped by his claim that at least half of the stars were orbited by planets. Just think; if half of the stars had planets, that was a lot of potential alien real estate.

In April that same year, Drake, unaware of Cocconi and Morrison's paper, had come to much the same conclusion concerning the microwave window and its suitability for alien communication. Audaciously, he had pointed the newly built 85ft radio telescope at the National Radio Observatory in Green Bank, West Virginia, at our two closest Sun-like stars, Tau Ceti and Epsilon Eridani, and listened. If those two stars had planets, and those planets were home to intelligent life, then maybe he might be able to pick up a signal.

Throughout the spring and summer of 1960, Drake had listened to the two stars for a total of 150 hours but found nothing. Nevertheless, a vital first step in humanity's communications with extraterrestrials had been taken. A fan of L. Frank Baum, he had named the project Ozma, after the Fairy Queen of the far-off land of Oz, and now other eminent scientists wanted to join him on the Yellow Brick Road. Not only would his hero Otto Struve be among the ten at Green Bank, but so too would neuroscientist John C. Lilly, famed for his work on dolphin communication;[7] Melvin Calvin, a chemist who would win a Nobel Prize for his work on photosynthesis on the first night of the conference; Philip Morrison, one of the two authors of the Cornell paper; and the gifted astronomer and adviser to NASA Carl Sagan.

Drake needed to set the pace for the conference, and – much in the mould of Fermi a decade earlier – decided the best way to do that would be to estimate the number of civilisations in the galaxy, N, that we might be able to receive signals from. The equation he came up with has become a landmark in the quest to communicate with alien intelligence, and here it is in all its glory:

$$N = R^* \times f_p \times n_e \times f_l \times f_i \times f_c \times L$$

Trust me, it looks a lot more ferocious than it is. And as a way of breaking it down, I want to take you to a Dire Straits gig.

BROTHERS IN ARMS

Specifically, I want to take you to Wembley Arena in the UK for the British leg of Dire Straits' famous 1985–6 world tour. Their latest album, *Brothers In Arms*, has gone

7 And about whom we'll be hearing more in our final chapter. Much more.

multi-platinum, and the 12,500-seater venue is packed to the rafters. The band close the show with their monster hit 'Money for Nothing'. Tired but happy, the army of fans make for the exit. But wait. The band come back onstage for an encore. Mark Knopfler strikes up the opening chords of the album's title track. Out in the corridors and turnstiles, the crowd hear that there is more to come, and about-turn.

Back in the auditorium, the seats start to fill. 'These mist-covered mountains . . . ' sings Mark. And one by one, they appear; cigarette lighters, held silently aloft. Human souls share the tender beauty of the song. Mark adjusts his sweatband and looks out into the darkness. It's a deeply affecting sight, but let's put our practical hats on for a moment. How many lit lighters does he see?

You might think it depends on how much of the song Mark plays before he looks up. Surely, you say, the number of lit lighters will slowly build up during the song as more fans enter the auditorium. If Mr Knopfler looks up after a minute, he'll see fewer lighters than if he looks up during the final few bars. But that's not necessarily the case.

It all depends on how long the lighters shine for. To see what I mean, let's imagine only one fan has a lighter, and that he can only keep it alight for thirty seconds before he burns his fingers. On average, will Mark see the light? Instinct will tell you he probably won't. For a start, he might look up before the fan has even entered. Or he might look up after the fan has entered, but after the light has gone out.

OK. So how many fans with lighters do we need to enter the auditorium during the four minute song in order for Mark to see at least one light? This is where maths comes in handy. Because to get the average number of lit lighters that Mark sees, we just need to multiply the rate at which fans with lighters enter the stadium by the length of time that they can keep them alight.[8] For example, let's say that eight fans with lighters enter the stadium during the song. In that case, the rate is 8/(4x60) fans per second, and the average number of lit lighters at any one time is

Rate × length of time alight = 8/(4×60) × 30 = 1

Interesting, eh? In other words, even if eight fans enter with lighters during the encore, on average when Mark looks up he will see only a single light.

MONEY FOR NOTHING

Right, now we know the principle of how this works, let's get down to some gritty detail. Obviously with a capacity crowd of 12,500 fans, our current figure of eight fans with lighters entering during the course of the song is way out. How can we more accurately estimate the true number?

There's going to have to be a bit of guesswork here. Could all of the fans make it back into the arena during the four-minute encore? Based on the recent Bill Bailey gig I attended, I'm going to say 'yes'. There will inevitably be a bit of a bottleneck at the entrances, but Wembley Arena is well served by generously proportioned walkways with exemplary signage. In fact, I am going to be so bold as to say that 10,000 of the fans make it back into the auditorium

8 In the case that there is only one obliging fan, the rate is of course 1 fan per 4 minutes, or 1/(4x60) fans per second. We already know that he can keep his lighter burning for a total of thirty seconds so we can calculate the average number of lit lighters Mark sees as (rate) × (length of time alight) = 1/(4x60) × 30 = 1/8. As this is less than 1, it means that on average Mark won't see a single lighter. Of course, if the scenario is repeated over the course of eight shows, we would expect that on one of those nights Mark would see a single light. I'm not sure how he would feel about that.

during the course of the four-minute song. That's a rate of 10,000/4×60 = 42 fans per second.

But wait. Not all of those fans will be carrying lighters. For a start, not all Dire Straits fans are smokers; in fact I would say quite the reverse; given the band's demographic, everyone attending is liable to be quite clean-living. Nevertheless, this is 1985, when menthol-flavoured cigarettes are considered to be a healthy option. Let's put the fraction that are smokers at 50 per cent. And, of course, not all of the smokers will be carrying a lighter. For the sake of illustration, let's say that 50 per cent of them are carrying two lighters apiece; their favourite Zippo, and a backup.

OK. So to estimate how many lighters Mark Knopfler sees, we need to multiply the rate at which fan-operated lighters enter the stadium by the length of time they are alight. In other words, if R is the rate at which the audience re-enter the arena, f_s is the fraction that are smokers, f_l is the fraction of smokers that have lighters, n_l is the number of lighters that each lighter-owning smoker carries, and L is the length of time in seconds that it is possible to hold a lighter without burning your fingers, then − big breath − the number of lit lighters Mark Knopfler sees during the encore is:

$$N = R \times f_s \times f_l \times n_l \times L$$

Let's plug in the numbers. In which case we get

$$N = 42 \times 50/100 \times 50/100 \times 2 \times 30 = 630$$

So on average, at any one time, Mark sees roughly 630 flames twinkling all around him in the darkened arena. He knows he is loved and no longer feels alone.

SO FAR AWAY FROM ME

As with pretty lights at an eighties stadium rock gig, of course, so with detectable alien civilisations. More or less, anyway. To figure out how many alien civilisations a radio astronomer might be able to see with his telescope, there are two things we need to get a handle on: the rate at which detectable alien civilisations emerge, and how long they are detectable for.

OK, so at what rate do detectable alien civilisations emerge? Let's assume that the kind of intelligent civilisation we are going to be able to communicate with is a lot like us; inhabiting a rocky planet in orbit around a Sun-like star. Then we can start with the rate of formation of Sun-like stars, R^*, and whittle it down just like we did in the case of the Dire Straits gig.

First, let's work on the rate at which Earth-like planets form. If R^* is the rate of formation of Sun-like stars, and f_p is the fraction of those stars with Earth-like planets, then their rate of formation is simply

$$R^* \times f_p$$

That's about as hard as it's going to get. Of course some solar systems will have more than one Earth-like planet, such as our own, where arguably the Earth, Moon, and Mars and maybe even Venus have been habitable to life at some point. This is a bit like our example of smoking Dire Straits fans who have more than one lighter, so if n_e is the number of Earth-like planets per solar system, then our running tally for the rate of formation of Earth-like planets is

$$R^* \times f_p \times n_e$$

Right. I'm sure you're getting the hang of this, so if f_l is the fraction of Earth-like planets that support life, the rate of formation of life-supporting planets is

$$R^* \times f_p \times n_e \times f_l$$

And if f_i is the fraction of life-supporting planets that have intelligence, the rate of formation of intelligent life is

$$R^* \times f_p \times n_e \times f_l \times f_i$$

We're so nearly there. We have found the rate at which intelligent life appears in the galaxy. All we need to do now is multiply by the fraction of alien intelligences f_c that have radio communication, and we will – at last – have the rate of formation of detectable alien civilisations. Here we go:

$$R^* \times f_p \times n_e \times f_l \times f_i \times f_c$$

Good. Now, just as in the case of cigarette lighters at a Dire Straits gig, the number of detectable alien civilisations will be equal to the rate at which they appear, multiplied by the length of time they last for. So let's put the cherry on top by multiplying by the length of time that an alien civilisation is detectable, L. And, lo and behold, we have derived the Drake Equation:

$$N = R^* \times f_p \times n_e \times f_l \times f_i \times f_c \times L$$

Satisfying, eh?[9]

---

9 Readers of my previous tome, *It's Not Rocket Science*, may spot that for the sake of keeping our relationship fresh I have used a slightly different version of the Drake Equation. Instead of R^* and L, that version uses N^*, the number of stars in the galaxy, and f_l, the fraction of a star's lifetime during which a civilisation is detectable. It all comes out in the wash, because $R^*=N^*/(\text{lifetime of a star})$ and $f_l = L/(\text{lifetime of a star})$. As the Americans say, 'you do the math'.

NARROWING THE ODDS

Back in the day, Drake and his colleagues at the first SETI conference judged the various factors of the Drake Equation to be as follows:

$R^* = 1$ (One Sun-like star forms per year)

$f_p = 0.2–0.5$ (Between a fifth and half of all Sun-like stars have planets)

$n_e = 1–5$ (Such stars will have between one and five planets in their habitable zone)

$f_l = 1$ (All such planets will develop life)

$f_i = 1$ (All planets that develop life will also develop intelligence)

$f_c = 0.1–0.2$ (Between 10 and 20 per cent will develop radio communication)

$L = 1,000–100,000,000$ years (The length of time for which signals are transmitted is between one thousand and one hundred million years)

Let's take all the lower limits:

$N = 1 \times 0.2 \times 1 \times 1 \times 1 \times 0.1 \times 1000 = 20$

In other words, on any given day, when we point our radio telescopes up into the sky there are twenty stars, spread throughout the galaxy, from which we might detect alien signals.[10]

Now let's take the upper limits:

$$N = 1 \times 0.5 \times 5 \times 1 \times 1 \times 0.2 \times 100,000,000 = 50,000,000$$

Meaning that there are fifty million stars from which we might pick up a signal. Simplifying things even further, we can see that, very roughly speaking, most of the factors are roughly 1 apart from L, the number of years that a civilisation is detectable.[11] We can then write the Drake Equation in a stunningly simple form, like so:

$$N \approx L$$

IS THERE ANYBODY OUT THERE?

Equations are like poems. There's what they seem to be about, and what they are really about. On the face of it, the Drake Equation simply tells us how to crunch the numbers to find out how many detectable alien civilisations might be out there in the galaxy, but of course there's something much more important going on. The real power of the equation is in the assumptions it forces us to make.

The deepest assumption is that the aliens will be just like us. We are presuming that the aliens will have technology like ours, societies like ours and planets like ours.

---

10 Or, rather, ten, since we can only see one hemisphere at a time.

11 In fact, Frank Drake himself has gone so far as to buy a personalised number plate with the letters "NEQSL", which is Geek for N = L.

Now I don't think for a second that Frank Drake is being naive; rather, his equation says, 'hey, we've got to start somewhere, so we may as well start here'. If it provides anything, the Drake Equation gives us a lower limit on what we might expect to find out there in the galaxy. After all, who's to say that aliens don't inhabit dust clouds in deep space as well as rocky metal-rich planets like our own?

The Drake Equation shows us that in considering the problem – namely, are we alone? – we need to think deeply about the very nature of life, intelligence, civilisation and technology. What do we mean by these things? And once we have made our assumptions, what data do we have that can turn them into bona fide estimates? For example, if we assume that biology is as universal as chemistry, how can our knowledge of how life evolved on Earth help us to make a guess about its abundance throughout the galaxy?

As you can imagine, these are some of the most fascinating questions a human mind can ponder. To find answers, we will be foraging on the very fringes of scientific knowledge. What do the latest telescopes tell us about the abundance of Earth-like planets? What do the latest advances in biology tell us about the nature of life, and the chance of it being commonplace in the cosmos? What is intelligence, and how might we communicate with an alien intelligence that is vastly different to our own? What do we want to say? And why do we want to say it?

LONG-DISTANCE RELATIONSHIP

So why do scientists believe in radio contact with alien civilisations, but not in flying saucers? On a simple level, you might say it's because of a lack of evidence. Given the fact that everyone now carries a mobile phone with a camera on it, you might think that there would be some really good footage of alien contact, but there isn't. What's more, no alien artefacts are on display in any of our museums, and no alien spaceship has landed on the White House lawn. What we do have is eyewitness reports.

One of the fun things about this book is that, in examining aliens, we really get to think about what it means to be human. And it is time to face one of the more unpalatable truths about our species: when it comes to the world around us, we apes are not the most reliable of witnesses.

As young children, we are bathed in imagination. We truly believe that we can fly, that we can see monsters in the wardrobe and that a fat, bearded Latvian man delivers all the world's Christmas presents on a fifteen-foot sleigh pulled by magic reindeer. We see a mixture of the world as it is, and the world as we imagine it to be. For children, wishing makes it so.

As adults, on the other hand, we pride ourselves on our impartiality. We are certain that the wild dreams that we have at night never intrude into our waking hours. We believe we see the world as it truly is, and consider ourselves the masters of our own imagination. By this logic, when an otherwise upstanding member of the community – a policeman, say, or a magistrate – sees a ghost in the middle of the night, his anecdote is all the proof that we need. Chris is a company director, Chris saw a ghost, therefore ghosts exist. But are our minds really as reliable as we think they are?

Your average scientist would say not. In fact you could say that one of the aims of science is to remove the so-called 'human factor' from our observations of the world; to try and describe the universe in an objective, logical, self-consistent way that can be tested by experiment. Thus evolution is taken to be true not because it's a great story and Darwin was a really steady guy, but because it predicted the existence of certain fossils before those fossils were ever found. Any given scientific theory stands only for so long as it is supported by experiment. It doesn't matter how many Nobel Prize-winning biologists believe in evolution; if a fossilised dolphin suddenly turns up among the trilobites in a piece of Cambrian sedimentary rock then we'll all be back to the drawing board.

Compelling as the stories about flying saucers are, and as much as I for one would like to believe that they are true, the evidence is of poor quality. Kenneth Arnold was, no doubt, a reliable man not given to exaggeration. He truly believed that he saw a fleet of strange craft that summer, flying across the snowline of Mount Rainier, and was as puzzled by what he saw as anyone else. As an amateur pilot, he had additional credibility; he had the skills to tell a fleet of spaceships from a flock of geese, for example, or from a formation of conventional aircraft.

Yet, harsh as it may seem, in scientific terms the reliability of Kenneth Arnold is neither here nor there. Science doesn't care who you are or what you think you saw, it simply demands evidence. You saw a flying saucer? Show me the footage on your smartphone. An alien spaceship crash-landed in New Mexico? Show me a piece of the ship. You were abducted by aliens and subjected to an internal examination? Show me . . . actually never mind.

'Ah,' I hear you say. 'But scientists are people, too. Why should I believe some hippy with a test tube over a model citizen like Kenneth Arnold?' And you'd be right. Trusting a scientist purely because of his or her name and reputation is a dangerous game. Scientists make all the mistakes that everyone else makes. Their imaginations play tricks on them, they cherry-pick data that supports their pet theories, and they have a biased view of their own talents and abilities. But experiment saves the day time and time again. For a scientific hypothesis to gain weight, it has to be testable by experiment, and that experiment has to be repeatable. Scientists do make mistakes, but every time they do experiment puts them back on the right track. Without hard evidence to back them up, despite seven decades of sightings, crash landings and abductions, the scientific case for alien artefacts is always going to be hard to make.

LITTLE GREEN MEN

Which brings us neatly back to Jocelyn Bell Burnell. Bell Burnell, you will recall, has picked up a series of pulses 1⅓ seconds apart, coming from within the constellation Vulpecula ('Little Fox') which is itself smack dab in the middle of the so-called 'Summer Triangle' of bright stars Deneb, Vega and Altair. She has a dilemma. No serious astronomy graduate wants to tell their supervisor they have intercepted an alien signal; on the other hand, no serious astronomy graduate doubts that as far as aliens are concerned, they are on the front line. If anyone's going to take the call, it's probably them.

Summoning all her courage, Bell Burnell telephoned her supervisor, Anthony Hewish, who was teaching in one of the undergraduate laboratories, and told him what she had seen. 'Must be man-made,' said Hewish, and came out to the telescope the next day to see the string of pulses for himself. Sure enough, there they were. He decided that there must be something wrong with the equipment, and for the next month he and Bell Burnell tried to eliminate as many sources of error as they could.

First, they confirmed that the source was keeping pace with the stars rather than with the Sun. Astronomers refer to this as keeping sidereal time.[12] That implied that the signal wasn't coming from Earth, ruling out man-made interference. Except, of course, that produced by other astronomers, who also keep sidereal time. Could it be that some neighbouring observatory was transmitting the signal as part of one of their research projects?

A letter from Hewish to all the neighbouring observatories drew a blank. What other straightforward explanations could there be? The team eliminated radar reflected off the Moon, signals from satellites and effects due to a large corrugated metal building just to the south of the telescope. They then checked all the wiring, which to Bell Burnell's considerable relief turned out to be sound. She had helped wire it, after all.

They made a thorough analysis of the pulses. The pulses were 1⅓ seconds apart, and each one lasted less than 0.016 seconds. That meant that whatever was producing them had to be small. Basic physics says that nothing travels faster than the speed of light, so the object producing them had to be, at most, five thousand kilometres across.[13] That's roughly the

12 A sidereal day is simply the time taken for the Earth to make a full rotation relative to the stars. If you were looking down on the Earth's circular orbit from above, you would see that by the time the Earth has made a full rotation relative to the stars, it has not quite made a full rotation relative to the Sun. To do that, it needs to turn for another 3 minutes and 56 seconds. What this meant for Bell Burnell was that if the source was located in the stars and appeared at 4.04 p.m. on one day, she would expect it to appear at 4.00 p.m. the next.

13 You can read all this in the original paper, 'Observation of a Rapidly Pulsating Radio Source' by A. Hewish, S. J. Bell, J. D. H. Pilkington, P. F. Scott and R. A. Collins, *Nature* 217, 709–13 (1968). At the time of writing you could view it for free on the *Nature* portal, http://www.nature.com/physics/looking-back/hewish/index.html.

size of the Earth (6,371km), which in astronomical terms is on the tidy side.

To try and get a handle on how far away the source was, the team measured the dispersion of the pulses. Dispersion, as you probably know, occurs when higher frequency waves travel faster through a medium than lower frequency ones. The classic example is of white light dispersing into all the colours of the rainbow as it passes through a glass prism. Interstellar space may not be full of glass, but it is far from empty. In fact, the spiral arms of our galaxy sit in a sort of gas of free electrons. Just as glass slows down low-frequency red light more than it does higher frequency blue light, this 'gas' of free electrons slows lower frequency radio waves more than the higher frequency ones. Send a pulse through interstellar space, and after a while, thanks to dispersion, the lower frequency parts of the pulse will start to lag behind the higher frequency parts. In the case of Bell Burnell's mystery object, by the time the pulses reached Earth there was a noticeable delay.

By measuring this time delay between the highest and lowest frequencies in the signal, and then using a simple model for the number of free electrons that the pulse had passed through, the team were able to work out how far away the source was. Their calculations placed it well outside the solar system but well within the galaxy, at a distance from Earth of roughly 200 light years.[14]

Next, they pondered what the set-up of this alien civilisation might be. What if the signal was coming from an

14 Once you start dealing with the enormous distances between stars in the galaxy, it gets a bit inconvenient to use puny Earth units like a metre. A much handier unit is the light year, being the distance that a beam of light travels in one year.

Earth-like planet which was in orbit around a Sun? What could they test? If the aliens were in orbit, surely their signal ought to be in orbit, too. If it was, sometimes it should be moving away from Bell Burnell's radio telescope, and sometimes it should be moving towards it. That movement should produce an effect; in fact it's a commonplace one called the Doppler Effect.

There's a classic example of the Doppler Effect, in the change in pitch of a passing train's horn. The train horn is producing one note. As it approaches, the sound of that note is pitched up. As it passes, the sound is pitched down. Put in more general terms, when the source of a wave signal is moving relative to a detector, it changes the frequency of that signal.

Was there a Doppler Effect in the signal? To the team's surprise, there was. Yet it wasn't due to the motion of the alien signal around some alien Sun. You see, Bell Burnell's telescope was itself in orbit, around our own Sun. The Doppler Effect that the team measured in the signal turned out to be due to the relative motion of the telescope to the source. As Bell Burnell herself wryly puts it, the team had simply managed to prove that the Earth revolves around the Sun. Reassuring, but of itself not much of a breakthrough.

They did make some progress. Once they had isolated this small Doppler Effect, they could see that the pulses in the signal were extraordinarily regular, with the gap between them varying by less than one part in ten million. That meant that whatever was producing them had a lot of mass, and therefore a lot of energy. If it was aliens, they really meant business. Someone had built themselves a very powerful transmitter indeed.

THE THIRD EYE

Feeling more and more certain that the signal they had found was not the result of some random man-made interference or faulty wiring, and increasingly sure that its source was something massive and extremely compact that was situated well within the galaxy but beyond the nearest stars, Jocelyn Bell Burnell and her team decided to bite the bullet. They would go and ask a rival telescope if they could see it, too.

Keeping it in the family, Hewish approached his colleague Paul Scott and his research student Robin Collins, who were operating a radio telescope at the same frequency. They calculated that the signal should show up in the second tele-scope just twenty minutes after it appeared in Bell Burnell's. As soon as the signal appeared in Bell Burnell's telescope, the team moved over to the second chart recorder. Twenty minutes went by with no signal. Hewish and Scott wan-dered off down the hall, with Bell Burnell tagging behind, discussing what could possibly cause the signal to appear in the one telescope and not the other. Suddenly, there was a shout from the lab. Robin Collins had hung back, waiting, and there the signal was, pulsing away. They had miscalcu-lated the delay by five minutes. The source was real.

On 21 December 1967, Anthony Hewish and the head of the group, Martin Ryle, held a meeting at the Mullard to discuss what to do about the object they half-jokingly called LGM, as an abbreviation for Little Green Men. If it really was a pulse from an alien civilisation, then those aliens were a contrary lot. For a start, the radio frequency of the pulse, 80MHz, seemed an unlikely choice; although perfect for quasars it happens to be a very noisy frequency.

If Bell Burnell's Little Green Man was signalling to Earth,

or other Earth-like planets, surely he would tune his signal to sit in the microwave window? Instead he had chosen a part of the spectrum where it would most likely get absorbed by gas and dust. Why send a signal at a frequency where it was less likely to be picked up?

Nevertheless, there the signal was, and if science teaches us anything it is to be humble in the face of the facts. If this really was an alien radio signal, who should they tell first? An astrophysical journal, or the Prime Minister?

Bell Burnell returned home for supper decidedly disgruntled. She had spent two precious years of her life wiring up a state-of-the-art radio telescope, ready to search for quasars. Instead, her experiment had been hijacked by a bunch of numbskull aliens. She had only got six months of grant money left and the window for her to finish her PhD and secure some sort of academic career was rapidly closing. As she put it herself, 'I was furious. For some reason, some silly lot of green men had decided to use my frequency and my aerial to signal to Earth.'

IN THE BLEAK MIDWINTER

That evening, Bell Burnell returned to the lab, determined to get back on track. A backlog of 2,500ft of chart paper had built up and was begging for analysis. Just before 10.00 p.m., when the lab was due to shut, she was looking at a section that belonged to the constellation of Cassiopeia when she thought she spotted some more 'scruff'.

Hurriedly, she laid out all the other bits of chart paper she could find that corresponded to Cassiopeia. There the 'scruff' was again. The timing couldn't be more acute. The next day she was going back home to Ireland for Christmas to announce her engagement. Calculating that the patch of sky she wanted would be in the telescope at around two o'clock in the morning, she decided on no sleep till Belfast, and headed over to the observatory.

This being the dead of winter, the equipment was cold and temperamental, but Bell Burnell 'breathed on it and swore at it, and I got it to work at full power for five minutes. It was the right five minutes and at the right setting. In came a stream of pulses, this time at intervals of one-and-a-quarter seconds, not one-and-a-third.'

That settled it. There was no fault with the equipment, no man-made interference; there was something out there in the stars. And it couldn't be Little Green Men. After all, what were the chances that there would be two lots of Little Green Men on opposite sides of the universe, both signalling at an obscure frequency to our little blue planet? Unlikely as it seemed, somewhere out there in the galaxy were massive, compact objects that produced pulses of radio waves. Jocelyn Bell Burnell had discovered the pulsar.

JOURNEY TO THE STARS

That might seem an anticlimax – we are hunting for aliens after all – but to my mind it's glorious. Null results might be the bane of pseudoscience, but they are a boon to science. No matter how much we want aliens to be out there, we have to go by the evidence. If SETI ever does pick up an alien radio signal, we can guarantee it will be subjected to the same kind of scrutiny that Bell Burnell's pulsar was, and that is a very good thing indeed. As the late Oliver Sacks put it: 'every act of perception is to some degree an act of creation, and every act of memory is to some degree an act of imagination.' When it comes to things we dearly want to believe, we have to be on our guard.

A pulsar is like an enormous lighthouse; a fast-spinning, highly magnetised ball mostly made up of densely packed material rich in neutrons, radiating a beam of electromagnetic radiation into space. They are the highly compressed corpses of large stars, formed after they have run out of fuel and exploded as supernovae. Each is unique, with its own distinctive type of radiation and pulse rate. Since Jocelyn Bell Burnell discovered them, we have found pulsars which spin so fast there are millisecond gaps between pulses. We have found still others whose 'beam' is made of x-rays, and others where it is visible light.

As we know, when NASA launched the *Voyager 1* probe, etched on to the casing of the Golden Record was a map showing the position of the Earth relative to its fourteen closest pulsars, with the pulse period of each pulsar coded in binary.[15] If ever an alien intelligence intercepts it and comes to pay us a visit, Bell Burnell's pulsar will be one of the landmarks that guides them here.

15 As I'm sure you'd want to know, the base unit is the hyperfine transition of hydrogen, the most common molecule in the universe. This is to do with something you may have heard of called 'spin'. Basically both the proton and electron in a hydrogen molecule have spin, and they like to line their spins up, either parallel or antiparallel. Sometimes an electron will emit a photon and flip from the higher energy state (spins parallel) to the lower energy state (spins antiparallel). The frequency of that photon – an extremely precise 1420.40575177MHz – can then be used to define a length of time. That's if the aliens don't tear the Golden Record off and eat it.

CHAPTER THREE

Planets

In which the author searches for Earth-like planets, learns about the Wow! Signal, and takes a stroll round Vienna with the UN Ambassador for the Human Race.

There is something oddly futuristic about the United Nations. Though the squat Arrivals building bears more than a passing resemblance to my low-rise 1970s primary school, the enormous courtyard I step out into is unfeasibly impressive. Everything about the place should seem dated: the mountains of grey concrete; the jet fountains that strafe an enormous shallow circular pool; the towering Cold War-style flagpoles; but instead the overall effect is of vertiginous progress. The trappings may all be mid-twentieth century, but the very existence of a super league of sovereign nations, united in the common interest of mankind, still seems like pure science fiction.

And it's this unique position in the world of human affairs that interests me today, because it's been widely reported in the British press that thanks to the recent discoveries of Earth-like planets by the Kepler Space Telescope – and the possibility that they might harbour intelligent life that we can make radio contact with – the UN is appointing a spokesperson for the human race. This 'Ambassador for Earth' has been named by no less a newspaper than the *Sunday Times* as one Dr Mazlan Othman of the United Nations Office for Outer Space Affairs (UNOOSA), and I have an appointment to meet her for lunch.

Yet as I mount the stairs to Dr Othman's office, the strong scientific imperative for my visit suddenly evaporates. This is the opposite of 'l'esprit d'escalier', a phrase nonexistent in French but which we English take to mean the inspiration which strikes as soon as an encounter is over and we are heading down the stairs on our way home. The more floors I climb, the drier my mouth gets and the sweatier my brow becomes, until all confidence in my mission has completely drained away.

In the world of extraterrestrial intelligence, this discomfort is commonplace, and is known simply as 'the giggle factor'. For some reason, when talking about the very real scientifically sound possibility of communicating with aliens, everyone gets the urge to laugh. And here, where national flags flutter at the tops of impossibly tall flagpoles, and where international diplomats negotiate the gravest of choices while pursuing the loftiest of ambitions, what on Earth do I think I'm doing asking the head of the UN's space executive about flying saucers?

It doesn't help that Dr Othman has an extremely impressive CV. Malaysian by birth and an astrophysicist by training, in the early noughties she spearheaded the Malaysian space programme, ANGKASA, and built a space observatory on the island of Langkawi, launched a remote-sensing satellite, *RazakSAT*, in the world's first near-equatorial Low

Earth Orbit, and oversaw the launch of the first Malaysian astronaut to the International Space Station in 2007. Since then, she has served as the Director of UNOOSA, and was appointed Deputy Director-General of the United Nations Vienna Office in 2009.

I needn't have worried. Once I have sweated and spluttered my way past her secretary in a manner even Hugh Grant would think was overegging it, Dr Othman greets me warmly, blaming the layout of the UN rather than my terrible sense of direction, and is disarmingly relaxed and informal. She leads me through to her office, a bright and breezy affair with a spectacular view across the Danube towards the Old City. Her desk sits in the far corner, half obscured by a jungle of luscious pot plants and, to my right, a sideboard displays glittering scale models of satellites and space stations.

We sit, and I do my best to try and convince her that I am not a crazy person, that I know my stuff about science, and that, while I think the evidence for UFOs is feeble, I am very interested in the possibility that there is intelligent, communicable life on other planets. I state my belief that biology is as universal as chemistry and physics, and that the recent discoveries of the Kepler Space Telescope have shown us plenty of places where that biology might get a chance to do its thing. In short, I do everything I can to try and reassure her that I am an emotionally well-balanced, scientifically literate individual with a passion for astrobiology. And, in doing so, I am fairly sure that I come across as a crazy person.

When I finally pause for breath, I see that the Director of the United Nations Office for Outer Space Affairs has a twinkle in her eye. 'Come on,' she says. 'Say it. You want to talk about aliens.'

TWINKLE TWINKLE LITTLE PULSAR

Strange as it now seems, as little as twenty years ago we still had no hard evidence that planets existed outside our own solar system. As an undergraduate student in the late eighties, I remember feeling very excited by reports that a planet had been discovered orbiting Gamma Cephei, a binary star some forty-five light years away in the northern constellation of Cepheus.[1] This seemed almost too good to be true. One of the most iconic moments of the first *Star Wars* movie had been Luke Skywalker looking out from his home planet, Tatooine, at two setting suns. Could it be that this new planet, like Tatooine, was a lawless desert world, blasted by the heat of two stars, where humanoid beings farmed moisture in underground dwellings? What else had George Lucas got right? Is it really possible to dodge a blast from a laser gun?

Sadly, though the first reports of this first planet were published in 1988, the year I began my PhD, they were then retracted in 1992, the year I turned professional as a comedian.[2] I only hope there was no connection. If there was, I needn't have worried, because that same year Aleksander Wolszczan and Dale Frail, working at the Arecibo Observatory in Puerto Rico, found the first bona fide planet in the constellation of Virgo. In fact, they found two, orbiting at roughly half the distance that the Earth orbits the Sun. Only they weren't orbiting a star. They were orbiting a pulsar.

1 The constellation of Cepheus is, of course, home to the famous Delta Cephei, a variable star whose brightness pulses with a repeating period of five days and nine hours.

2 There's a happy ending, by the way; the planet Gamma Cephei Ab was finally confirmed in 2002.

That, obviously, was not what we were expecting at all, and I doubt that Frank Drake was straining at the leash to point a radio telescope at PSR B1257+12 to try and pick up a message. For a start, it's a thousand light years away so the conversation would be a little stilted. And, secondly, pulsars are awesome things, but for life as we know it, being blasted by x-rays is always going to have its drawbacks.

Then finally, in 1995, we discovered 51 Pegasi b, the first planet orbiting a Sun-like star. That wasn't quite what we were expecting either. Fifty light years away in the constellation of Pegasus, it was half the mass of Jupiter, but extremely close to its home star, taking only 4.2 days to complete an orbit. How had it got so close? After all, in our own solar system it's small rocky planets like Mercury, Venus, Earth and Mars that sit close to the Sun. Gas giants like Jupiter and Saturn roam much further out, and medium-sized icy stuff like Uranus and Neptune sit out in the boondocks. Could it be possible that, in some solar systems, planets didn't stay put?

Once someone had found something that was definitely a planet, the floodgates opened. Radio astronomers everywhere tried to get in on the act, and all manner of planets turned up. Many of them, like 51 Pegasi b, were so-called 'Hot Jupiters': large gas planets tucked up close to their home stars. Others were so-called 'Hot Neptunes': medium-sized planets that had also migrated into tight orbits. Still others were truly monstrous creations patrolling at unfeasibly large distances.[3] What we didn't find was anything like the Earth.

Which is to say we found nothing Earth-sized that was

---

3 2M1207b, for example, was found to be over three times the mass of Jupiter, with an orbit eight times wider.

an Earth-type distance from a Sun-like star. One of the things we think makes Earth such a good home for life is the fact that most of it is covered in water. In fact, astronomers define something called the habitable zone, which means the range of orbits where the temperature of an orbiting planet is going to be neither so hot that water simply vaporises (such as on Mercury and Venus) nor so cold that it freezes (such as on Mars). In the official jargon, we couldn't find any Earth-sized planets in the habitable zone of their home star. Could it be that, far from being mediocre, the Earth was incredibly special?

Formally, this became known as the Rare Earth Hypothesis. Microbial life might be common, the argument went, but intelligent life is rare because planets like the Earth are rare. After all, many things conspire to make the Earth ideal for life. Firstly, it's a good size. Much smaller, and its gravity would be too weak to hold on to an atmosphere, and without a decent atmosphere there would be no greenhouse effect to keep the surface warm. This, of course, is the problem with Mars, which has only been able to hold on to the thinnest of atmospheres.

Secondly, it's volcanic. As we shall shortly see, one of the most promising hypotheses for how life got started depends on volcanic springs, and, in any case, the gases released and consumed by the rock cycle play a vital role in maintaining a life-friendly carbon dioxide-rich atmosphere. Not only that, but plate tectonics also ensures that heavier elements like metals get recycled into the oceans and atmosphere, providing lots of esoteric chemistry that life can make use of.

Thirdly, the Earth has a strong magnetic field. Not only is that handy if you are trying to find your way around, but it means that we are protected from the hail of damaging radiation known as the solar wind. Fourthly, it has a large Moon, which not only slows its rotation, giving milder weather, but keeps the Earth's axis pointing in the same direction. Without a Moon, the Earth's spin axis might flip, wreaking climatic havoc. And, finally, it has Jupiter as a cosmic vacuum cleaner to protect it from comets.[4]

One look at the Chicxulub crater in the Yucatán Peninsula will tell you that comets are bad news. That particular impact wiped out the dinosaurs, and we have Jupiter to thank for the fact that such Armageddons are relatively rare events. Basically, the bully planet's strong gravity hoovers up anything dodgy that the outer solar system throws our way. Incredibly, in 1993 we actually saw this in action when the comet Shoemaker–Levy was spotted orbiting Jupiter, disappearing in a spectacular collision just over a year later.[5]

In truth, none of these arguments have really gone away. Intelligent life may be rare. After all, if our nearest neighbours are in the next galaxy, rather than in the next star system, that goes a long way towards explaining the Fermi Paradox. Nevertheless, we have some cause for hope. Not everyone believes that a flip-flopping north pole would be a disaster, for example; and there are arguments that, while Jupiter protects us from comets, it wreaks havoc in the asteroid belt. But by far the most encouraging evidence has come from the Kepler Space Telescope; in short, the size and orbit of the Earth aren't nearly as rare as we might have feared. In fact, they are decidedly run-of-the-mill.

---

4 Loosely speaking, a comet is an icy rock from either the Kuiper Belt – the doughnut-shaped ring of icy rubble that is home to Pluto, and currently being charted by NASA's New Horizons probe – or the Oort Cloud.

5 We believe Shoemaker-Levy started life in orbit around the Sun, but was captured by Jupiter some two or three decades before it was discovered.

PLAYING WITH A LOADED DECK

It turns out that back in the nineties and noughties, the method we had for finding planets relied on them being big. Called the radial velocity method, it basically relied on detecting planets via gravitational effects. As a large planet orbits a star, it causes that star to wobble, and the wobble affects the frequency of the light that the star gives out. If you analyse the way that the star's light changes frequency you can work out how fast it is wobbling, and once you know that you can then work out the mass of the planet and the size of its orbit.[6]

All very clever, but the only planets you can detect tend to be large ones. Small rocky planets like the Earth don't cause their home stars to wobble, or at least the wobble is so small that it hardly affects the star's light and so is very hard to measure. To find out whether there were any Earth-like planets out there, we needed a new telescope. And as of 2009, that's where NASA's Kepler mission came in.

THE MAN IN THE MIRROR

It's truly fitting that the Kepler Space Telescope is named after Johannes Kepler, the seventeenth-century German astronomer. Not only did he invent the modern refracting

---

6 What I'm describing, of course, is another example of the Doppler effect. When a wave source moves relative to an observer, it affects the frequency that the observer receives. The classic example is the increase in pitch of a train's horn as it approaches, and the drop in pitch when it passes.

telescope,[7] beloved by amateur astronomers the world over, but he also placed an important foundation stone upon which we built our understanding of planetary motion. Using observations made by the Danish astronomer Tycho Brahe, Kepler proposed that the planets' orbits aren't circular, as had been believed since the time of Plato, but elliptical. What's more, he claimed that the planets' speeds aren't constant, as had also been believed since the time of the Ancient Greeks, but vary according to where they are in their orbit. One of the crowning achievements of Newton's *Principia* was that he was able to show that his Law of Universal Gravitation, when applied to the Sun and planets, produced all the effects proposed by Kepler.

The optical telescope that bears Kepler's name is a spectacular beast. As the name suggests, it is far from earthbound; it was launched into space on a three-stage rocket into a heliocentric, or Sun-centred, orbit. The advantages of being up in space are several: for a start, the atmosphere blurs starlight, which is why the stars twinkle, and, secondly, you don't have to wait for it to get dark to be able to use it. The business end points north, so that it never catches the Sun, and looks at a small area of the sky on the edge of the constellation of Cygnus, the Swan. I say small; within that area it monitors the brightness of some 145,000 stars.

Kepler uses what's called a transit method to detect

7 OK, I can't let that go without a few qualifications. Galileo is usually credited with the invention of the refracting telescope, but what he actually did was improve on the 1608 design of the Dutch eyeglass maker Hans Lippershey. Kepler's trick was to use two convex lenses, rather than one convex objective lens and one concave eyepiece lens as used by Lippershey. It's Kepler's 1611 modification that forms the basis of contemporary refracting telescopes, though other advances have also been made such as use of achromatic lenses (lenses which treat all colours of light the same).

planets. Put simply, it looks at the light coming from a given star, and if it sees a dip in brightness, it knows a planet is crossing, or, to use the jargon, transiting. Needless to say, it's very sensitive, as it's looking for something like the drop in brightness of a fruit fly passing across a car headlight. By measuring exactly how much light is blocked, it can work out how big the planet is, and by measuring the time between transits it can work out how large the orbit is and how hot the planet must be.

The results have been astonishing. To date, Kepler has turned up over one thousand planets, with over three thousand prime suspects awaiting confirmation. And what have we discovered? Well, hindsight is a wonderful thing, but in many ways it's obvious: big and small planets are less common than medium-sized ones. In other words, most planets appear to be between Earth and Neptune in size. Smaller planets like Earth, and big ones like Jupiter are plentiful, but not quite so abundant.

As a result, we now believe our own solar system is something of an outlier. For a start, Jupiter-sized planets are unusual, and tend to be hot rather than cold like ours. Secondly, we don't have that many medium-sized planets. And, lastly, there's nothing inside Mercury's orbit, whereas many of the systems found by Kepler have planets of all kinds in orbits of ten days or less.[8] That said, the solar system's not exactly rare: we currently think the odds of a cold Jupiter are something like one

8 The so-called Grand Tack model links these three facts: Jupiter and Saturn formed closer to the Sun than they are now, at roughly 3 AU (1 AU is the radius of Earth's orbit). The remaining gas in the disk slowed Jupiter down, and it migrated towards the Sun, ejecting the material that would have formed super-Earths. When Jupiter reached roughly 1.5 AU, the gravitational pull of Saturn forced it to change direction. Eventually Jupiter came to rest in its present orbit of 5 AU, with Saturn at 9 AU.

in a hundred. With 200 billion stars in the Milky Way, that still leaves around a billion planetary systems similar to our own.

When it comes to Earth-like planets, the Kepler data is also encouraging. The latest estimate is that over one in five Sun-like stars have an Earth-sized planet in their habitable zones, implying the nearest wet rocky planet might be as little as twelve light years away.[9] Although Kepler stopped making measurements in 2013, we are still sifting through the mountains of data it produced, and finding more and more small rocky planets in the habitable zones of stars like our very own Sun.

The problem with Kepler, of course, is that to maximise its chances of success it was pointed towards a distant but dense clump of stars. That means that all the Earth-like planets we have found so far are well out of the range of even our most advanced telescopes. Kepler 186f, for example,[10] is a rocky planet roughly the same size as the Earth orbiting its home star at the right distance to have water on its surface. All right, that home star happens to be an M-type rather than a G-type like our own Sun – a 'red dwarf' to you and me – and such stars are prone to scorching solar flares.[11] But Kepler 186f happens to be at the outer edge of its star's

9 Tantalisingly, there is some evidence that Tau Ceti, the thirty-fifth most distant star from the Sun and our nearest lone Sun-like star, may have an Earth-sized planet in its habitable zone. Tau Ceti just so happens to be twelve light years away. It's also one of the two stars that Frank Drake targeted in Project Ozma.

10 Discovered 7 April 2014. There's an article on NASA's website here: http://www.nasa.gov/ames/kepler/nasas-kepler-discovers-first-earth-size-planet-in-the-habitable-zone-of-another-star/

11 Stars are classified by colour, from blue to red, large to small, O B A F G K M, producing one of the most questionable mnemonics in science, 'Oh Be A Fine Girl Kiss Me'. The Sun is a G-type star. Stars become increasingly abundant as you move from O to M. Roughly 8 per cent of stars are G-type, 12 per cent are K-type and 77 per cent are M-type. All the other classes make up just 3 per cent of the total.

habitable zone, just out of harm's way. And that means it may be suitable for life.

SETI, of course, wasted no time in pointing a radio telescope at Kepler 186f, and searching up and down the dial for anything that looked suspicious. They found nothing. That doesn't mean there were no radio signals being transmitted, because Kepler 186f is nearly 500 light years away, and to be detectable any transmitter on Kepler 186f would have to be ten times the strength of the Arecibo Radio Telescope here on Earth. Or to put it another way, if a civilisation like ours exists on Kepler 186f, the SETI search wouldn't have found it. There might be sentient beings on Kepler 186f right now, uploading the secrets of the universe on to the intergalactic internet; we'll never know.

Happily, NASA's next generation space telescope, the Transiting Exoplanet Survey Satellite (TESS), launches in 2017 and will pick up where Kepler left off, surveying half a million of our nearest stars and hopefully pinpointing thousands of Earth-like planets. No doubt SETI will be quick to target them, but so too will other next-generation space telescopes like the James Webb. Designed to work in the infra-red, this fabulous piece of kit will be perfect for analysing planetary atmospheres. After all, molecules like oxygen, carbon dioxide and nitrogen have a distinctive 'bar code', emitting and absorbing light at well-defined frequencies across the spectrum. If we know what we're looking for, there's every chance we may be able to detect alien life remotely. Our next-door neighbours may have their lights off and their curtains closed, but telescopes like the James Webb can tell us whether or not they are home. Then Frank Drake will know exactly where to point his telescope.

THE RESTAURANT AT THE BEGINNING OF THE UNIVERSE

As we walk to lunch Mazlan tells me how the media came to refer to her as the Alien Ambassador. 'I was due to give a talk at a Royal Society conference about extraterrestrial life. I was going to say that if we do receive signals, the United Nations is the best way to coordinate a response.'

We pass UNOOSA's space display, and I am temporarily distracted. There's a beautiful model of the *Shenzhou* space-craft, which will be the shuttle craft for the Chinese Space Station, together with its *Long March* launcher rocket. There is something vaguely familiar about both – the technology is essentially Russian, after all – but it is remarkable to see Chinese characters on the side of spaceships. What a different world it will be in the 2020s, with the International Space Station decommissioned and only Taikonauts in orbit.

Most of the display items are scale models of satellites; one of UNOOSA's tasks is to provide a registry of all items launched into space, and then help keep track of anything that ends up in orbit. There are many reasons to love satellites, and one of them has been early warnings of climate change: their data have given us 90 per cent confidence that the planet is warming due to carbon dioxide emissions. That said, what most people love about satellites is the money they make. With GPS, television and the internet all relying on satellite transmissions, virtually every nation on the planet is trying to get a piece of the action.

Only that morning I had read an article in the *Daily Mail*, the gist of which was that while we were pouring aid into Nigeria, they were squandering it on a space race. After feeling suitably furious, it struck me that Nigeria's space race was probably little to do with planting the Nigerian flag on Mars and more to do with satellites. After all, launching satellites is probably the most sensible way of supplying infrastructure to a developing country that you can possibly imagine. I checked online, and surprise, surprise, that is indeed the purpose of Nigeria's space programme, with telecommunications and Earth observation satellites bringing internet services, weather-mapping and food security to one of the fastest growing populations on the planet.

And, joy of joys, among the display items there's also a moon rock, found by the astronaut James Irwin of *Apollo 15* on the rim of the Spur Crater in the Mare Imbrium, better known as the right eye of the Man in the Moon.[12] The *mare*, or seas, on the Moon are basically enormous impact basins, formed by collisions with asteroids or comets. These basins then flooded with molten lava, which cooled to form huge flat plains of dark basalt, making them ideal landing sites for the early Apollo missions. It's a sobering thought that the Earth has been similarly disfigured throughout history, though of course thanks to weathering and recycling of its crust via plate tectonics its impact basins have been Botox-ed away.

'So there's nothing in it?' I ask. She smiles, and we continue our walk. 'A journalist called after the story first broke. She asked me, "Are you the alien ambassador?" I said, "I have to deny it. But it sounds pretty cool."'

When it comes to messaging aliens, of course, the UN has got form. As we heard in the opening chapter, it's Secretary-General Kurt Waldheim's voice that opens *Voyager*'s Golden

---

12 To be clear, I am taking his right eye to be the large dark circle on the top left as we look at him, handsome chap. By the way, if you haven't already done so, I heartily recommend downloading the latest version of Google Earth and taking a trip to the Moon.

Record, and I think I know an audition speech when I hear one. The way things are going, very soon the peoples of Earth are going to need someone to speak on their behalf. Isn't this the role the UN was born to play? Someone must have thought this through. If the aliens call, surely somewhere among all those reports and resolutions there has to be a protocol? Dr Othman laughs. 'Here at the UN, we simply serve. We don't create protocols unless we are mandated to by our member states.'

Suddenly it hits me. There's only one thing worse than the aliens talking to the UN, and that's them talking to just about anyone else. After all, we kind of know how this goes. In 1996, when American scientists in Antarctica thought they had found fossilised bacteria in a Mars meteorite, the first the rest of the world knew about it was when President Clinton announced it on TV.[13] We need to keep politicians out of it; they'll just hog the glory. The last thing any of us want to see is a humanoid alien in the Downing Street rose garden, hand in hand with David Cameron. That business with Nick Clegg was close enough.

It's time to put Dr Othman on the spot. What if an alien ship lands tomorrow? I wince, expecting her to tell me not to be so silly. To my great surprise, she hardly breaks stride.

'It depends where they land. If they land in Mali, they will be the provenance of Mali.'

'Really? But what about the UN?'

'If the government of Mali requested that we became involved, we would get involved.'

'And if they did make that request?'

'Then we would need to get it verified. We could help

13 More on the so-called Allan Hills meteorite in Chapter Five.

assemble a team of scientists, and assist in obtaining visas, but that could take a couple of months.'

'But SETI have a protocol, don't they, for what to do if a ship lands?'

'There's a SETI protocol, sure. But it has never been adopted [by the UN].[14] There has never even been a debate.'

WILD MEN OF THE MOUNTAINS

The Search for Extra-Terrestrial Intelligence has always struggled to get taken seriously. First of all there's that acronym, 'SETI', which sounds a bit too close to 'yeti' for comfort, and immediately puts the reader in mind of the Bigfoot hoaxes involving blurred camerawork and out-of-work actors blundering about in bad costumes made of 1970s shagpile carpet. And then there's the Steven Spielberg movie *ET*, in which an alien lands on Earth in a spaceship, for ever more linking the word 'extraterrestrial' with UFOs. After all, it's hard to make a serious argument for SETI when your mental image of an alien contact is a wrinkly brown baby-faced midget with a glowing forefinger.

This mistrust is completely undeserved, if you ask me, and not a little unfair. An admittedly subjective sample of the public, based on taxi drivers, people I've sat next to at weddings, and fellow travellers on Network SouthEast rail tells me that people take UFOs very seriously indeed, but are – initially, at least – extremely dismissive of SETI. I can't

14 Sorry, I've always wanted to do that square brackets thing. The UN's position on all things extraterrestrial is to be found in the Outer Space Treaty, a copy of which you can find on the UNOOSA website, here: http://www.unoosa.org/oosa/SpaceLaw/outerspt.html

understand it. SETI is conducted by professional astronomers on state-of-the-art radio telescopes, while the search for UFOs is conducted by drunk people on their way home from the pub. The science behind UFOs is nonexistent; the science behind SETI is sound. And SETI had its greatest champion in the gifted astronomer Carl Sagan; UFOs have their strongest advocates in the shape of the Church of Scientology. I rest my case.

As a result of the public's ambivalence, SETI has been a bit of a stop-start affair. The project began with two independent events. Let's have a quick refresher on what those were. Firstly, in 1959 two Cornell physicists, Giuseppe Cocconi and Philip Morrison, published a paper in the journal *Nature*, where they pointed out that the microwave radio band would be a good way for extraterrestrial civilisations to communicate with us, because shorter wavelengths tend to get absorbed by the Earth's atmosphere and longer ones by the gases in the interstellar medium. They put forward the idea that within this band there was one obvious marker frequency emitted by neutral hydrogen, the most common molecule in the universe.[15]

Frank Drake had independently come to the same conclusion, and in 1960 pointed the large telescope at Green Bank at our two closest Sun-like stars, Tau Ceti and Epsilon Eridani, and tuned his receiver to look for signals close to 1,420MHz. As his receiver had a bandwidth of 100Hz, in the technical jargon we might say that he searched one 'channel' of 100Hz. He found nothing, and thereby expanded our knowledge of extraterrestrial

---

15 The hyperfine transition of hydrogen, used as a measurement of time on the Golden Record. The corresponding '21cm hydrogen line' has a precise wavelength of 21.10611405413cm in free space, which corresponds to a precise frequency of 1420.40575177MHz.

civilisations: there wasn't one on either of those two stars. Or, to be more precise, he found no extraterrestrial civilisations that were sending us radio signals close to the hyperfine transition of neutral hydrogen for the 200 hours that he listened in April 1960.

After Frank Drake's promising start, the US initiative in SETI failed to attract US government funding and stalled. Drake and others continued to beg, borrow and steal time on radio telescopes whenever they could, but NASA was slow to take up the cause. In the Soviet Union, however, where the scientific community had not been weaned on *War of the Worlds* and *John Carter of Mars*, SETI seemed like less of a joke, and more like a subject for serious scientific study. So after a promising start in the West, with the 1959 paper by Cocconi and Morrison, closely followed by Frank Drake's vigilante Project Ozma, most of the running on the theoretical side of things was made by the Soviet Union.

Interestingly, the Soviet take was very different from the American one. Whereas the Americans took it for granted that they would be able to recognise an alien signal, the Soviets weren't nearly so sure. After all, Ancient Egyptian had only been translated with the help of the Rosetta Stone. Even then it had taken twenty years of academic slog, and Ancient Egyptian was a human language. What hope did we have of recognising an alien signal even if we found one?

RUSSIAN DOLLS

Instead, the Soviets focused on something much more fundamental: energy. Drake's counterpart in Soviet SETI is the Russian astrophysicist Nikolai Kardashev. In his 1963 paper 'Transmission of Information by Extraterrestrial Civilizations', Kardashev classified civilisations by their energy consumption. It was statistically likely, he argued, that most civilisations in the galaxy are much older than our own. In other words, the chances are we've just rocked up at a party that's been swinging for billions of years.

The older a technological civilisation is, argued Kardashev, the more advanced it would be, and therefore the more energy it would require. He decided to classify them as one of three types, solely on the basis of their energy consumption. A Type I civilisation was one which had harnessed all the energy of its home planet. By this measure, our own civilisation isn't quite a Type I, but it's not far off. A Type II civilisation had command of all the energy of its home star; a Type III civilisation had harnessed the energy output of its home galaxy.

Kardashev's paper said little about what these civilisations would actually be like, but did give a nod to a contemporary paper by the British-born American physicist Freeman Dyson, who had proposed a kind of Type II civilisation which has come to be called a Dyson sphere. The gist of Dyson's 1960 paper is this: an advanced extraterrestrial civilisation might completely surround its home star with a swarm of artificial structures, to make use of every scrap of light. These structures would effectively enclose the star, completely blocking out most forms of electromagnetic radiation.

Objects like the Kepler Space Telescope, which orbits the Sun, are, of course, the first bricks in a Dyson-sphere-like wall. But if an advanced civilisation completely surrounded its home star with light-blocking satellites, how would it show up in our telescopes? Dyson's answer was heat. Heat, of course, is just another part of the electromagnetic spectrum, known as the infra-red. So in principle, these sorts of civilisations should show up in infra-red space telescopes if they are sensitive enough. Look for a large object radiating a lot of heat but little visible light and you just might find ET.[16]

So while the Americans were searching nearby Sun-like stars for microwave radio signals near the hydrogen line, the Russians were scouring the skies looking for large objects radiating in the infra-red. Of the two strategies, you'd have to say that the Russians seemed to have the edge. A Type II civilisation might be difficult to spot with a ground-based infra-red telescope, which was all that was available at the time, but what about a Type III civilisation? Who could fail to spot something the size of a half-blotted-out galaxy, where the dark bits were radiating strongly in the infrared? The Americans, on the other hand, were searching for a needle in a haystack. And yet it was the Americans who found something first.

THAT'S, LIKE, WOW!

I hope that got your attention. Because if you think that SETI has so far drawn a complete blank, you need to know about the Wow! Signal. It's a bittersweet story, because unfortunately it's the alien version of a kiss on the cheek rather than a committed long-term relationship. But again, I urge you; what follows is not the dramatic story of a UFO 'enthusiast', but a measurement that a very real SETI scientist made with a very real scientific instrument.

16 Bloody hell, I'm doing it now. Needless to say, I'm not talking about a wrinkly brown alien with a glowing finger, I mean extraterrestrials in general.

That scientist was the astronomer Jerry Ehman, and the instrument was the Big Ear Radio Observatory, a now defunct radio telescope belonging to Ohio State University. In fact, it's more than defunct: it's an eighteen-hole golf course. The Big Ear was a Kraus-type radio telescope, meaning it wasn't the dish type you may be more familiar with, but instead consisted of two huge rectangular reflectors at either end of an enormous aluminium sheet the size of three US football fields. One reflector was flat, the other curved, and they were set up so that incoming radio waves bounced off the flat reflector, on to the curved reflector, and then into a detector. In fact, it's such an ingenious set-up I think it's worth drawing, so here it is:

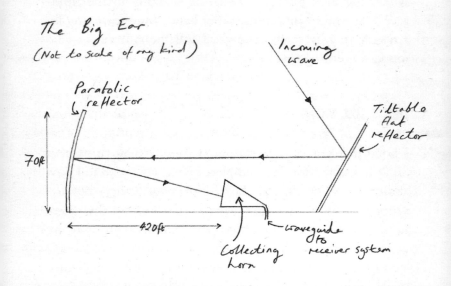

The Big Ear started its working life in 1965, and for twelve years it provided the Earth's most detailed mapping of cosmic radio sources, known as the Ohio Sky Survey.

Jerry Ehman had been a big part of that project, joining in 1967 after completing a PhD in astronomy at the University of Michigan. Sadly, in 1972 the Sky Survey had its funding withdrawn and Jerry lost his job. Unable to secure another research post at Ohio State, he started teaching business classes at nearby Franklin University while continuing to work on the Big Ear as a volunteer.

Rather than see their state-of-the-art telescope go to waste, Jerry and others relaunched it as a SETI project. Together, they converted the Big Ear from what's known as a wideband instrument to a narrowband one. Most natural radio sources are wideband, meaning that they give off electromagnetic waves with a wide band of frequencies, sometimes from high-frequency x-rays all the way out to low-frequency radio waves. A good example would be quasars. Narrowband sources – no prizes for this – are bunched into a narrow band of frequencies. They are nearly always artificial; the best example would be a radio station.

The flat reflector at the Big Ear would be set into position, then as the Earth rotated it would scan a strip of the entire sky. After twenty-four hours, the angle of the reflector would be minutely adjusted, and the next strip would be scanned. Jerry and his ex-colleagues set up the Big Ear's computer so that it would sample the intensity of the signal arriving in each of its fifty channels, convert it to a number between 0 and 35, then print it out.[17] Or, rather, a number between blank and Z, because space on the printout was limited and that way two-digit numbers could be printed as a single digit: 10 was A, 11 was B and so on.

17 The observant among you will notice that fifty channels was a big improvement on Frank Drake's single-channel set-up. In other words, the Big Ear was searching a much wider band of frequencies around the hydrogen line than Frank's telescope at Green Bank had been sixteen years earlier. What's more, it was searching them remotely.

On Friday, 19 August 1977, Jerry Ehman sat down at his kitchen table to go through the latest batch of printouts. There on the right-hand side of the printout were the coordinates of the patch of sky that the Big Ear had been looking at, together with the time it had made each observation. And down the left-hand side were the signal intensity readings: the usual motley crew of blanks, 1s and 2s as the telescope soaked up the silence of the spheres. But wait. There, in channel 2, was something extraordinary: 6EQUJ5. Ehman took his red pen and circled the six digits, then next to them scrawled a single word: 'Wow!'

So what did those digits mean? To you or me they might make an excellent personalised number plate for the presenter of a popular gameshow,[18] but to Jerry Ehman they described a signal of unusual intensity, some thirty times more intense than the cosmic background radiation, flashing through the detector. Let me convert it to asterisks for you, so you get a feel for it.

Usually, Jerry would see this:

*

*

**

*

But this time he saw this:

18 I am more than ready to admit that's a minority interest gag. I was going for 6E QUJ5 = SEXY QUIZ. *Pointless,* maybe?

```
******
*************
**************************
******************************
*******************
*****
```

And just to remind you, this compares to a maximum possible intensity of Z, or

```
***********************************
```

After a bit of number-crunching, Jerry was able to determine that the rising-and-falling shape of the signal was due to the movement of the telescope rather than any change in the intensity of the broadcast, which had remained more or less constant. The detection had lasted only seventy-two seconds, but it was impossible to say with certainty whether the source had been broadcasting for days, months or even years before that because, as luck would have it, this was the first time that the Big Ear had scanned that particular strip of sky looking for narrowband signals. It was also impossible to tell whether or not the signal was carrying information; it could have been AM, FM, a single frequency switching rapidly or slowly on and off, or not varying at all.[19]

The most logical explanation seemed to be that an earthbound artificial signal had been bounced back into the telescope somehow, yet all efforts to explain Wow! in this way foundered. There had been no planets in the right

19 Quick refresher: for a signal to carry a substantial amount of information, it has to vary. For AM signals, the information is coded into changes in the signal's amplitude; for FM signals it's coded into changes in frequency.

position to reflect Earth signals back into the telescope, neither had there been any large asteroids or satellites. Aircraft, spacecraft and ground-based transmitters were also ruled out. There was no getting away from it, the detected signal was exactly what you would expect from a point source located in the furthermost stars. What's more, it was confined to a very narrow band of frequencies near the hydrogen line, just as predicted by Cocconi and Morrison all those years before.

Working back from the position data on the printout, Jerry and his team were able to narrow down the source of the signal to two small patches of sky in the constellation of Sagittarius.[20] As you may know, from our point of view here on Earth, Sagittarius hangs like a net curtain over the centre of the Milky Way Galaxy, and behind it lies a clot of distant stars. The same strip of sky was scanned again the following night, but the Big Ear heard nothing. In fact, the team kept the Big Ear in the same position for the next sixty days and nights, but the signal never returned.

So was the Wow! Signal the first message received by mankind from an extraterrestrial civilisation? Possibly. For us to be certain, we'd have needed to be able to repeat the observation, but so far all attempts to pick up more signals from that particular corner of the constellation of Sagittarius have proved fruitless. But it's interesting, eh? Suddenly, the idea that we might pick up a signal in the very near future doesn't seem so crazy after all . . .

---

20 Two patches, because the Big Ear had two receiving horns at the focus of its parabolic reflector. The signal had only entered one horn before it cut out, but it was impossible to tell which one. Two horns implied two slightly different patches of sky.

THE REAL JODIE FOSTER

Jill Tarter certainly doesn't think so. Jill is the astronomer that Jodie Foster's character in the movie *Contact* is based on.[21] From 1995 to 2004 she ran Project Phoenix, a SETI search of our nearest Sun-like stars. As ever with SETI, Phoenix relied on private funding, and had no dedicated radio telescope. Instead, it hopped around the globe, from the Parkes Observatory in New South Wales,[22] up to the National Radio Astronomy Observatory, Green Bank, West Virginia, and finally to the Arecibo Radio Telescope in Puerto Rico. Rather than scan the whole sky like Big Ear, Phoenix targeted 800 of our nearest Sun-like stars, listening in on billions of narrow channels across a wide range of radio frequencies.

Sadly, at the end of this impressive search the result was nil: not one solitary alien signal. Does this mean we are alone? Not at all. As Jill Tarter put it in a recent interview, 'the haystack we are searching in is vast'. In fact, let's put some numbers to this. Even if our galaxy contains 10,000 communicable civilisations among 200 billion stars, we'd need to search around ten million stars before we found anything. So far we have searched a paltry 10,000 stars, and we've only really searched for one kind of signal; the kind we would send if it was 1950 and all we had was radio telescopes. So the silence doesn't mean there's nothing out there, it just means we live in the sticks. Who knows? As far

21 If you haven't seen it, I suggest you put that right immediately. Based on the novel by Carl Sagan, Jodie Foster plays a SETI researcher who picks up an alien signal and Matthew McConaughey. And the most believable of those two things is the aliens.

22 Famous for being the main receiving antenna for the *Apollo 11* moonwalk, as exaggerated to great comic effect in the movie *The Dish*.

as the Intergalactic Federation is concerned, maybe we live in a conservation area.

Clearly there is more work to be done, and thanks to a generous donation from the co-founder of Microsoft, Paul Allen, since the late noughties SETI has had its own purpose-built radio telescope. The Allen Telescope Array at Hat Creek Radio Observatory in Northern California represents something of a departure from the norm. Instead of building one large dish, the feeds from a large number of smaller six-metre dishes are combined using sophisticated electronics to create a much more versatile instrument, able to survey a wide area of the sky at a wide range of radio frequencies with extremely high resolution. In its current incarnation, the ATA has forty-two dishes; later stages will see it grow to 350 dishes.

The ATA has three very exciting goals. Firstly, it is going to search the nearest million stars for artificial signals over a large chunk of the microwave window, from 1 to 10GHz. So if there are any alien civilisations broadcasting radio waves within 1,000 light years of the Earth, we are going to find them. Secondly, it is going to do the same thing for all the exoplanets that the Kepler Space Telescope has discovered. And thirdly – get this – it is going to survey the forty billion stars of the inner Galactic Plane for strong artificial signals near the hydrogen line. I don't know about you, but I think it's time the UN got itself a protocol.

FOLLOW THE METHANE

Back at the UN, Mazlan Othman and I have finished our lunch. In fact, we've been talking so long that the waitresses are wondering whether to offer us another meal. We've somehow got on to the topic of what a highly evolved species might look like, and whether they would even be that interested in the physical universe at all.[23] As Mazlan puts it, why go to the trouble of building a spaceship to visit new worlds when you can just build virtual ones? That's one answer to the Fermi Paradox: the aliens aren't here because they are playing nine-dimensional *Tetris*.

The dream of SETI is to receive an extraterrestrial message that we can understand and respond to. But would it ever really be possible to communicate with species much more evolved than our own? 'You and I need this,' says Mazlan, knocking on the tabletop. 'We rely on the physical world for our existence. When I no longer have need of the material world, what am I? Maybe the purest form that a being can take is energy.'

I start to worry that someone might overhear us and call security. Isn't she worried that this kind of loose talk is somehow, you know, unprofessional? 'I think it's a missed opportunity when people don't want to talk about aliens. I want to hear what people believe. If it engages people, draws them in, then what have you got to lose? Because then we can ask ourselves, "what's true?" And then the discussion becomes meaningful.'

So what will it take for one of the member states to bring the SETI debate to the UN? What are the hard facts that would close the case for communicable extraterrestrial civilisations? A signal, obviously, would be pretty conclusive. But what about other evidence, like the number of potential habitats? After all, the Kepler Space Telescope has shown us that there are at least as many planets in the Milky Way

23 That's certainly the case with my nine-year-old son, who spends more time building virtual worlds in Minecraft than he does constructing dens in the garden.

Galaxy as there are stars. What's more, those planets aren't just found orbiting Sun-like stars, they are everywhere: orbiting red dwarf stars, binary stars, pulsars and even wandering rogue, answering to no one. With so many places that life could take hold, surely the case for other intelligent technologically advanced civilisations is easily made? What do we need to do to get people's attention?

'You need to find life. Bacterial life.' Does she mean on Mars? After all, NASA's *Curiosity* rover has proved that Mars was once habitable. 'Maybe. Though Mars is so close to Earth that life could easily have travelled between them.' She's got a point. As we shall see in the next chapter, evidence is growing that Mars, not Earth, had the chemical catalysts necessary to kick-start life. Once microbial life got going on Mars, the theory goes, it was then carried to the Earth by a meteorite. If that does turn out to be the case – and that's a very big if – Martian life will turn out to be a branch on the same family tree as Earth life, and we'll all be back to the drawing board. So where else should we be looking?

'I find Titan exciting. There's a methane ocean, with lots of carbon, and that's what we're made of. If we found bacterial life on Titan, we'd really be making the case that life is widespread.' Titan, you'll remember, is the largest moon of Saturn, with sandy beaches lapped by methane oceans, and methane clouds scudding across a sky pumped full of nitrogen. Could it be that a completely independent type of microbial life has made its home there?

And if Titan fails us, there's always the icy moons of Jupiter. Beneath a crust of ice, Europa has an ocean of liquid water, and an atmosphere rich in oxygen. What's more, unlike Titan, which currently has no scheduled mission, Europa is going to get a fly-by when the European Space

Agency's *JUICE* probe launches in 2022. Maybe there are microbes lurking deep down in Europa's hydrothermal vents? Curiosity may have proved that Mars was once habitable, but Europa is habitable right now.

THE HITCHHIKER'S GUIDE TO VIENNA

The art of being a good visitor is knowing when to leave, and every fibre of my being tells me that time was several hours ago. Mazlan asks me what my other business is in Vienna that day, and I give her the honest answer: to buy an ice cream and wander around the Old City. How about her? It is then that my genial host confesses that today is her day off. I am truly impressed. There are few people who would willingly spend time with me when paid, let alone when they could be watching daytime television with their feet up. As she is leaving the office, she says, how about she points me in the right direction, as the Old City is on her way home?

What follows is a tram ride round old Vienna, taking in the Opera House, the Austrian Parliament and the Natural History Museum, though not necessarily in that order. With my bearings well and truly set, we then wander through the cool green of the Volksgarten, alone but for four silent American tourists riding on Segways. The other side of the Volksgarten gives out on to the Heldenplatz, and there looming in front of us is what looks like four Buckingham Palaces tongue-and-grooved together.

This, I learn, is the Hofburg Palace, one-time stronghold of the Habsburgs, and even today the official residence of the Austrian President. It is also home to the National Library, which is what Mazlan wants to show me. We cross the Heldenplatz, past two fierce statues of military heroes on rearing horses.[24] Usually it wouldn't strike me as at all odd to celebrate war and learning in the same monument, but in my current mindset I can't help but wonder how all this would come across to the crew of an alien star cruiser.

Mazlan leads me into the library, where we find a riot of marble pillars, detailed frescoes and Ancient Greek statuary. And it's then that I have that day's second lightbulb moment, only this time it's of the slightly less embarrassing, insightful kind. And that insight is this: culture is key. In the Drake Equation, it's just another number, the length of time for which a civilisation is detectable. But culture is more than just the glue that holds a civilisation together. Once I can write something down in a book, I no longer have to remember it. Statues tell a story that survives long after the heroes they commemorate have passed on. Culture is our way of storing information outside ourselves; a leap from the physical to the virtual.

I look up at the vaulted ceiling of the Hofburg to find an enormous biblical fresco, and I remember something else Mazlan had said, back when we were discussing what highly evolved beings might be like: 'Maybe God is the alien. He is in all of our minds. He is pure consciousness.'

24 Later I learn that these are Prince Eugene of Savoy and Archduke Charles of Austria, two of the most successful military commanders in Austrian history. It was Prince Eugene who defeated the Turks on behalf of the Habsburgs in the seventeenth century, while Archduke Charles – himself a Habsburg – defeated Napoleon in 1809.

CHAPTER FOUR

Universes

In which the author brushes up on his astrophysics, builds a pocket cosmos and discovers that if the universe had been created almost any other way then none of us would be here to witness it.

So a picture is emerging. Since the middle of the last century, the search for aliens has been divided into two camps: the UFO enthusiasts, who definitely aren't scientists; and SETI, who definitely are. No prizes for guessing whose side I'm on. The irony, for me, is that so far the UFO lot have had the best press. Most punters, I would hazard, have heard of Roswell while few have heard of the Wow! Signal. So far, SETI may have drawn a blank, but given the paltry number of stars we've searched, that's exactly what you would expect. And thanks to NASA's Kepler Space Telescope, in the future we'll have a much better idea of where to look.

As the discovery of Kepler 186f makes clear, when it comes to SETI it's all about timing. Kepler 186f is 500 light years away, so for a signal to reach us today not only would it have to have been sent by a transmitter ten times more powerful than any we have on Earth, it would have to have been sent 500 years ago. We know that communicating civilisations are possible, because we are here. But to detect our neighbours, they need to transmit a signal at just the right moment in the past for us to intercept it in the present. All this is expressed in the Drake Equation, of course, but there's nothing like a real-world example to bring it into focus.

It's stating the obvious, but the reverse is also true. To message our neighbours, we need to allow our signals time to cross the galaxy. And if we want those signals to reach millions of planets, they need to be strong. Back in the day, our TV was broadcast by powerful transmitters. *Fawlty Towers*, for example, which first aired in 1975, is only just reaching alien worlds thirty-five light years away. *The Office*, on the other hand, may never arrive. By that point, we were watching on cable, satellite or DVD, none of which leak much of a signal into space. If you think about it, as far as radio transmissions go, we were only noisy for a handful of decades.[1]

And even if we are lucky enough to catch some neighbouring aliens during their equivalent of our twentieth century, there's no point striking up a conversation if they live a thousand light years away. No signal – be it radio, laser or modulated cosmic ray – can ever travel faster than the speed of light, so each time we sent a message, we'd have to wait 2,000 years before we heard anything back. Even if Frank Drake's guess is correct, and the average civilisation lasts 10,000 years, we'd just about have time for

1 If we ever pick up a stray transmission from an alien world, my guess is that it will be multi-camera with a studio audience.

'Hello, what's the secret of life?' ... 'Thank you very much, goodnight.'

To be truly communicable, our neighbours need to be close by. Luckily, they may well be. As we've learned, the Kepler data has shown us that over a fifth of Sun-like stars have Earth-like planets, implying the nearest one might be as little as twelve light years away. Spookily close, eh? We've been scouring the distant galaxy, and they were in our backyard all along. Maybe our nearest M-type star, Proxima Centauri, is home to an Earth-like planet? If that's the case, then they probably can pick up *The Office*. In fact, one or two of them may even be in it.[2]

Kepler has given us a good feel for the fraction of Sun-like stars that have Earth-like planets, all of which is grist to the mathematical mill that is the Drake Equation. Now we need to know the fraction of Earth-like planets where life blooms. Thankfully, the extraordinary story of how life on this planet got its start provides vital clues as to what we can expect to find out there in the galaxy. Once again, I have to say you are reading this book at just the right time. Our understanding of Earth-based life has been making extraordinary progress over the past few years; in fact, to such an extent that many prominent astrobiologists are beginning to wonder whether it really is Earth-based at all.

CRAZY LITTLE THING CALLED LIFE

Like all great endeavours, the story of life on Earth is basically a game of two halves. In the first half, a microscopic

2 I always did think the sudden appearance of Ricky Gervais was suspicious.

universe emerges in a Big Bang of impossibly dense energy, then rapidly inflates, then settles down into a slow expansion, eventually cooling to form stars and galaxies.[3] In the second, the Earth forms, develops an atmosphere and is bombarded by water-laden comets which create the oceans. Life appears, first as single-celled microbes, then becomes multicellular,[4] then evolves into complex organisms like you and me. And here's the rub. In the first half, the game is rigged in life's favour. In the second, everything is left to chance.

To see what I mean, let's look at the first half of that story. It's often said that the Earth is a 'Goldilocks' planet, because, just like Mummy Bear's porridge, it is neither too hot nor too cold, but just right. 'Just right' in the case of the Earth means at the right distance from the Sun to have oceans of liquid water, and in fact we talk about a 'habitable zone' as being the band of orbits that a planet can sit in where it is neither too close to its home star so that water vaporises nor so far away that it freezes.

But it's not just the Earth that is incredibly well suited to life. Kick the tyres of creation and almost everything you find seems rigged in life's favour. Consider this: the age of the universe, its mass, the rate at which it's expanding, its lumpiness and the relative strength of its four fundamental forces are all incredibly finely tuned. Change any one of them slightly and life would cease to exist. Forget about a Goldilocks planet, we live in a Goldilocks universe.

---

3 Actually, for the last billion years or so the expansion has been speeding up, due to a mysterious thing we call dark energy. More on this shortly.

4 An organism is multicellular if it is made up of more than one cell stuck together; a chain of bacterial cells would fit the bill, but a colony wouldn't. Multicellularity has evolved independently nearly fifty times, and has been around for at least 1.5 billion years. Complex life, on the other hand – the special kind of multicellular life in which cells are grouped into tissues, and tissues into organs – has been around for roughly 575 million years. Arguably, complex life has evolved six times: in red algae, brown algae, green algae, fungi, plants, and animals.

A LITTLE HISTORY OF THE UNIVERSE

Let's just remind ourselves of what an extraordinary place the universe is, by brushing up on our basic cosmology. Firstly, no one knows its true size. It may be infinite, it may be one of an infinite number, we don't know. One thing is certain, however: it has a finite age. The latest data, from Europe's Planck satellite, clocks the cosmos at 13.8 billion years. For comparison, the age of our home galaxy, the Milky Way, is 13.2 billion years; the Earth is a mere stripling at 4.55 billion years.

The fact that the universe has a finite age means that there's a limit to how far out we can see with our telescopes, because this is also the age of the oldest light. At the time of writing, one of the most distant object known is the galaxy UDFj-39546284, which we are seeing a mere 480 million years after the Big Bang.[5] The light from UDFj-39546284 has therefore taken over thirteen billion years to reach us, which is even longer than the average pharmacist takes to locate your medicine after you hand them your prescription.

A finite age, of course, also implies that the universe had a beginning. We call that beginning the 'Big Bang', a wry phrase coined by the twentieth-century British astronomer Fred Hoyle, who was himself a proponent of the rival 'Steady State' theory. Despite his gentle mockery,[6] it is the Big Bang which has proved the test of time. The universe is

5 Discovered in January 2011 by the Hubble Space Telescope. UDFj-39546284 is contained within a famous image called the Hubble Ultra Deep Field, a tiny window of space in the southern constellation of Fornax.

6 I have to point out that Hoyle himself claimed no ill feeling towards the idea of an expanding universe. You be the judge.

expanding, as we can readily see from the red-shifted light of distant galaxies and supernovae; if we rewind the footage that means it must have originated from a single point. Poetically, cosmologists call such a point a *singularity*, because it operates outside the known laws of physics.

Just so as you know, the universe has expanded in four distinct phases. Immediately after the Big Bang came an extremely short, slow, steady expansion. The second phase, called inflation, was extremely rapid, lasting only from 10^{-36} to 10^{-32} of a second.[7] The third, lasting roughly thirteen billion years, was also slow and steady. The fourth, which has lasted roughly a billion years, has seen the universe's expansion speed up, as if some repulsive force is finally winning out over gravity.[8] This is the epoch in which we have the great fortune to find ourselves.

Some people worry about the Sun running out of fuel, which is predicted to happen in about five billion years' time. Not me. I worry about the runaway expansion of the universe. The fact that the furthermost galaxies are moving faster and faster away from us means that, in a few billion years or so, they will be moving away from us at the speed of light. At that point, they will disappear from our telescopes, leaving only darkness. Eventually, of course, all but the very closest galaxies will do the same, and one by one the stars will go out, too. Tell me that doesn't give you the most horrendous claustrophobia. If the Sun dies, we can just move to another star, but jumping ship to another universe is a whole other matter.

We have a name for whatever it is that's causing this

---

7 A word of warning; some take the words 'the Big Bang' and 'inflation' to mean the same thing. Not so here. Within these pages, 'the Big Bang' means the beginning of the universe, and inflation is what happens next. You've got to have a system.

8 The evidence comes from distant supernovae, which we can see are accelerating away from us.

runaway expansion of space: dark energy. Many cosmologists believe it's a property of space itself, called vacuum energy. In essence, the idea is that space is never really empty, but contains a swarm of so-called virtual particles. As the name suggests, they don't stick around for long. An electron and a positron, for example, will spontaneously appear out of nothingness, have their fun, then annihilate one another within a vanishingly small fraction of a second. Cheeky as this sounds, it's permitted by quantum theory, where you can borrow any amount of energy provided you borrow it for a short enough time. The snag is that the rate of expansion caused by all the virtual particles in the universe should be much, much greater than that we see. In fact, it should be about 10^{120} times greater, a discrepancy which is known fondly in the cosmology community as 'the worst prediction in physics'.

Secondly, we aren't entirely sure what most of the universe is made of. In a nutshell, we can see that nearby galaxies are spinning too fast to be held together just by normal matter in the form of stars, gas and dust; we call the missing stuff 'dark matter'. Dark matter, whatever it may be, exerts a gravitational pull on ordinary matter, but little else. Our best guess is that it is some sort of as-yet-undiscovered particle with no charge and a large mass. The leading candidate is a theoretical particle called a neutralino, though, despite a worldwide effort by thousands of scientists running an impressive array of detectors, no one has yet managed to track one down.[9]

Whatever dark matter is, it's definitely the boss of you.

---

9 If you've heard of supersymmetry, you may be interested to know that the neutralino is the supersymmetric partner of the weak gauge boson. Like other gauge bosons, such as the newly discovered Higgs particle, the neutralino is also its own antiparticle.

Incredibly, the ordinary matter of the periodic table makes up only 15 per cent of the total matter in the universe; the other 85 per cent is dark. There are some outlying theorists who suggest that dark matter is made up not of one but a whole zoo of particles, and is therefore also capable of forming dark black holes, dark planets and possibly even dark life, but given that we have yet to find even one dark matter particle, the jury's out.

And thirdly, despite an unpromising start in a chaotic molten fireball, both dark and ordinary matter is full of glorious structure. Like wheels within wheels, planets orbit stars, which are grouped into galaxies, which club together to form galaxy clusters, which in turn federate into galactic filaments. Lumpiness now implies lumpiness earlier, and in fact one of the great triumphs of experimental cosmology has been to trace the structure of stars and galaxies right back to the fireball, only 380,000 years after the Big Bang.

That needs a word or two of explanation. You'll remember that the microwave background radiation is the remnant of the light from the Big Bang, released when neutral atoms formed around 380,000 years after inflation. At the time it was set free it was extremely high-energy – think of x-rays and gamma-rays – but as the universe expanded, it cooled, becoming microwaves.

Since the microwave background radiation was last in thermal equilibrium with the rest of the universe when neutral atoms formed, it should bear the imprint of any lumpiness in the form of temperature variations. Yet back in October 1988, when I started my cosmology course, every effort to find any temperature variation in the microwave background had returned a negative result. If the lumpiness was there, it was too fine-grained for our best telescopes to spot.

It was NASA's COBE satellite, launched in November 1989, that finally found what we were looking for. The temperature variations in the cosmic microwave background radiation were tiny, but they were there, winning the COBE team the Nobel Prize in Physics in 2006. At last we knew: the fireball had not been smooth, it had been lumpy. What's more, the lumpiness had a specific value of 1 part in 100,000, as predicted by our theories of inflation.[10]

As always with cosmology, there's a catch. Inflation theory predicts that on a scale greater than that of filaments there's no more lumpiness, a phenomenon that cosmologists rather grandly call 'The End of Greatness'. Interestingly, no one seems to have told the universe. As we improve our telescopes, bigger and bigger stuff seems to turn up.

The record used to be held by something called the Great Sloan Wall, a filament of galaxies a billion light years across. Recently, however, astronomers found the first stirrings of something roughly ten times larger, called the Hercules–Corona Borealis Great Wall, or HCB GW for short. And it's not an isolated result. The latest data from the Planck satellite seems to show that there is more matter in one half of the universe than the other, which implies there are even bigger objects out there yet to be discovered. It's beginning to feel like every time we increase the volume of space we survey, we find an even larger structure.

Yet impressive as such objects undoubtedly are, it's at the scale of galaxies that the universe holds a truly numinous beauty. Our own discus-shaped Milky Way is a stunning example. A giant swirl of hydrogen gas, dust and stars

---

10 The basic idea is that quantum fluctuations in density in the microscopic universe immediately after the Big Bang were amplified by inflation.

circulating an enormous central supermassive black hole, within its spiral arms stars of all sizes are constantly forming as thick clouds of gas and dust collapse under their own gravity.[11] And it's within stars that atoms are made.

The bigger the star, of course, the hotter and faster it burns, and the bigger the atoms it can make. Smaller stars like our own Sun are capable of making lighter elements, but the biggest stars can make every element in the periodic table, right the way up to uranium. At the end of their lives these giant thermonuclear pressure cookers then explode in what is called a supernova, showering the surrounding gas with freshly minted atoms.

If this aforementioned gas then collapses under gravity to form a new star, some of these atoms then find themselves clumping together under gravity to form asteroids, moons and planets. Our own Sun is just such an example, burning hydrogen to make helium, surrounded by a belt of rocky planets, then a rocky asteroid belt, then giant gaseous planets, then giant ice planets. It's extraordinary to think it, but all stuff that surrounds the Sun is a remnant of ancient supernovae. You, this book and this planet are literally made from stardust.

A POCKET UNIVERSE

All very well, you might think, but what has all this got to do with life? What difference would it make if the universe was an infinite number of years old rather than 13.8 billion?

---

11 Actually it's only a spiral as far as ordinary matter goes. If we could only see dark matter, it would look more like a globe, with the supermassive black hole at its centre, and the glorious bright spiral of the Milky Way stretched across the equator.

If all we experience from dark matter is the pull of its gravity, couldn't we do without it? Does it really matter how big galaxies are, or whether stars make atoms?

Fascinatingly, the answers are: a lot; not in the slightest; yes, enormously; and you'd better believe it. To see what I mean, we are going to create our very own model universe, and then start messing with it. And if you are willing to accept that the very fact that you are reading this book requires stable stars, an abundance of hydrogen, carbon, nitrogen, oxygen and maybe a little phosphorus, and enough time for evolution to take place . . . well, then, you are going to be in for a bit of a shock.

GIVE IT A WHIRL

To get a taste of what I mean, think of that most mysterious of forces, gravity. In daily life it feels like a hindrance, pinning us to our beds in the morning, exhausting us at the gym and flinging each and every delicious slice of toast on to the kitchen floor butter-side down. One of the great joys of the Moon landings was seeing NASA's astronauts roam free, leaping across the lunar surface in giant strides despite being mummified in water-cooled suits and carrying backpacks the size of wardrobes.

Yet compared to the other forces of nature, gravity is extraordinarily weak. A hydrogen atom, for example, is made up of a sole proton orbited by a single electron. There are two forces at work: the electromagnetic force, which attracts the negative charge on the electron to the positive charge on the proton; and the gravitational force, which attracts the heavier proton to the lighter electron. Staggeringly, the gravitational force is roughly 10^{36} times weaker than the electromagnetic force. If gravity is so feeble, perhaps we could do without it altogether?

But consider this: without gravity, there would be no life anywhere. Why? Because there would be no universe. Since the middle of the twentieth century we've known that all creation began in a Big Bang, inflating out from a minuscule fireball only a fraction of the size of an atom, and cooling into the colossal mix of ordinary matter, dark matter and dark energy that we see today. In other words, though it started with nothing, today the universe contains a great deal of energy, present in several different forms.

So where did all this energy come from? How do you get something from nothing? Gravity is our best answer. In the broadest of terms, we think that the 'positive energy' of all the stuff in the universe is balanced by the 'negative energy' of its gravity. The total energy of the universe is therefore zero. For this reason, the originator of inflation theory, the American physicist Alan Guth, has described the universe as the 'ultimate free lunch'.

So without borrowing energy in the form of gravity, the universe would never have got its start. But that, of itself, doesn't dictate how strong gravity needs to be to produce a universe that is capable of bearing life. And here's the really interesting thing. If gravity were appreciably stronger, life would simply not exist. Intrigued? Then read on, because, quite frankly, that's not even the half of it.

WORKS STRAIGHT OUT OF THE BOX

OK, let's play this game. Let's imagine we are superbeings and we have been given our very own universe for Christmas. It's basically an empty glass box about half a metre across, sitting on a wooden base. It comes with a row of dials, a green button that resets to empty and a red button that initiates the Big Bang. No need to look for batteries; it doesn't need any, because the universe comes for free.

To begin with, the dials are all pre-set. If we push the red start button, we see the universe emerge in a bright Big Bang, then immediately inflate, then expand and cool to form stars and galaxies. If we are really keen-eyed, we'll notice that the expansion speeds up over the last second of its 13.8 second life-time.[12] The whole thing then freezes, preserving the universe as it is at the present moment. If we switch the lights out in our superdwelling, we can see the whole observable universe, with the Virgo supercluster at the centre. And though it's much too small to see when viewed on this small scale, at the centre of the Virgo supercluster is the Sun, orbited by the Earth.

Right, so let's find the gravity dial. It's there on the far left, next to the dials for the other three fundamental forces: electromagnetism, the weak force and the strong force.[13] Straight away, we can see that something's up. The dial for the strong force is on a linear scale, and set to 1, but the dials for the other three forces work in powers of ten. The electromagnetic force is set to −2, the weak force to −6, and gravity at a whopping −38.[14] What happens if we reset with the green button, reduce gravity by a few clicks then hit the red button to restart the universe?

Amazingly, we see something even more grand and

12 In other words, as superbeings we experience a billion years as exactly one second.

13 Of the four, gravity and electromagnetism are the only two we ever really witness in our everyday lives. The weak and strong forces are much more short-range, and hold domain over the world of the atom.

14 Gravity is therefore 10^{36} times weaker than electromagnetism, as mentioned earlier.

life-friendly. We see the tiny flash of the Big Bang, followed by inflation, just as before. The first stars take a split second longer to form, but when they do they are truly gargantuan. Once again they go supernova, spreading the elements of the periodic table throughout their neighbourhoods. Galaxies also form a little more slowly, grow larger and the stars within them burn for longer. This is especially handy for stable hydrogen-burning stars like our Sun. With three clicks down on the dial, to −41, such stars last for much longer, giving evolution time to do its thing time and time again.

In other words, as far as life goes, lower gravity is good news. But what about an increase? Let's go up four clicks, to −34, hit the red button and see what happens. Fascinatingly, as far as life is concerned, it's a disaster. This time it takes much less matter to form stars, planets, galaxies and black holes. In comparison to the normal universe, galaxies are tiny and form much more quickly, with small stars so close-packed that they are constantly stealing one another's planets. And the average lifetime of a star like the Sun decreases, lasting only hundredths of a second. Even if such a star manages to hold on to planets, there won't be nearly enough time for life to emerge. Strong gravity means no life.

KNOTS IN COTTON

OK, so we've toyed with gravity. What happens if we tweak the lumpiness of the early universe? Let's find out by locating another handy dial, Q, that controls the density variations in the fireball. Like the dials for the electromagnetic, weak and gravitational forces, it's on a logarithmic scale, set to −5.[15]

15 In other words, such that the fireball density varies by 1 part in 100,000.

Let's dial up the magnification on our model, light up the dark matter and hit the slo-mo button, so we can see exactly how this process works.

Firstly, if we really zoom in, we can see that the density variations start life as quantum fluctuations, which then get amplified as the universe is inflated. Over the next 0.4 of a second – corresponding to some 400 million years in the real universe – pockets of dark matter form, growing denser and denser, which hydrogen and helium then fall into. Eventually the ordinary matter becomes so dense that it collapses under its own gravity, lighting up as the universe's very first generation of stars.

And some stars they are, too. Some are hundreds of times the mass of our own Sun, and when they explode as supernovae they shower their own particular corner of the universe with every chemical element imaginable. What's left of their cores then collapses down to form a black hole, and continues to pull a swirl of matter and dark matter towards it. As the black hole starts to become supermassive, that swirl becomes a galaxy.[16] And galaxies are the nursery beds for generations of stars and planets.

AVENUE Q

OK, so that's the universe we've already won. That's safe. Now let's do what superbeings were put on this superplanet

16 Cosmologists call these first stars the Population III stars. Obviously it would have made more sense to name them the Population I stars, but the stellar populations are named in the order that they were discovered rather than the order in which they arrived on the scene. I say discovered; we have found Population I stars (young and metal-rich, e.g. the Sun) and Population II stars (old and metal-poor, e.g. the stars in the bulge at the centre of the Milky Way) but for the time being Population III stars are theoretical only; everyone has their fingers crossed that one of them turns up in the James Webb.

to do, and tweak the lumpiness of the primordial fireball to see how that affects the evolution of the universe. Let's start by dialling Q down, to 1 part in 1,000,000, and see what effect it has.

The difference is striking. After we press the red button, we see the flash of the Big Bang, then inflation, then darkness. The fireball is too fine-grained for any part of it to collapse under gravity. The expansion wins out, and once again we end up with a thin gruel of hydrogen, helium and dark matter particles. There are no stars, no planets and no galaxies. Once again, the universe is a no-go zone for life.

What about increasing Q? Surely that will make glorious giant galaxies, like when we reduced gravity? Sadly not. If we dial Q up to −4, reset and push start, before a second is up, huge regions of dark matter begin to clump together, precipitating enormous clumps of gas, more massive than entire galaxy clusters, that quickly collapse under gravity to form truly forbidding black holes. That's not a reality any of us want to live in.

THE STRONG FORCE

By now, I'm sure you're getting the picture, but before we move on there's one more tyre I want to kick: the strong force. Without it, of course, there would be no atoms, because its job is to bind protons together to make atomic nuclei. You'll remember that this happens in two places: firstly, in the Big Bang, where the nuclei of lighter elements like lithium and helium get made; and secondly, in big stars, which make all the rest.

That sounds pretty straightforward, and you might think that so long as the strong force is attractive, and can be felt when protons get within a short range of one another, all would be well. But there's a complication. Protons also carry a positive charge which tends to push them apart. At the same time that the strong force pulls together all the protons in a nucleus, the electromagnetic force is pushing them apart. In fact, one of the puzzles of the early twentieth century was how atomic nuclei managed to exist at all.

'Very well,' you might say, 'let's just make the strong force between two protons in a nucleus stronger than the electromagnetic force between them.' But that's not how the problem is solved either. As it turns out, the strong force isn't quite up to the job of holding two protons together in a nucleus; instead, the electromagnetic force wins and the two protons are forced apart. So how do atomic nuclei remain stable?

The solution, ingeniously, comes in the form of a neutral particle with roughly the same mass as a proton, which also feels the strong force, but doesn't have any charge. No doubt you'll remember that such particles are called neutrons. These gentle peacemakers are the diplomatic glue that holds atomic nuclei together. They also feel the strong force, but they don't have any charge, so aren't repelled by the positive charge on the proton. A deuterium nucleus, for example, which is made up of a proton and a neutron, is stable, as is a helium nucleus, which has two protons and two neutrons.[17]

17 You'll remember that we call the number of protons in a nucleus the atomic number, Z. The total number of protons and neutrons is called the mass number, A. Atoms with the same number of protons but different numbers of neutrons are called isotopes. Protium (P), deuterium (D) and tritium (T), for example, are all stable isotopes of hydrogen with zero, one and two neutrons respectively. As a shorthand, we tend to let the name of the element stand in for its atomic number, and specify the isotope using the mass number. ^{3}He therefore has two protons and one neutron; ^{4}He has two protons and two neutrons.

FAITES VOS JEUX

Right. With all that in mind, let's twiddle the dial for the strong force on our model universe and see what happens. First of all, let's dial it up a bit. Surely, if we increase the attraction between protons and neutrons, making nuclei will get easier? And with more elements knocking around the periodic table, there will be even more chemistry and hence a greater abundance of life?

I'm afraid not. Let's increase the strong force by a tenth, to 1.1 on the dial, and press the red button. Once again, we see the flash of the Big Bang, and half a second later the Dark Ages end, as the first giant stars burst into life and go supernova. Galaxies form just as before, but we notice something odd about the stars within them. They have exceedingly short lifetimes; in the compressed time of our model universe, they last maybe a millisecond or so. Once again, there's not nearly long enough for evolution to produce life. And even worse, if we zoom in on the asteroids and planets, we see that none of them have any water. What's going on?

It was all there in the Big Bang, but we missed it. Crucially, increasing the strong force has meant that all the protons fused into pairs, forming a new helium isotope with no neutrons, ^{2}He. In other words, we have made a universe with no hydrogen. It's the hydrogen burning phase in stars that takes time, so they burn much quicker. And without hydrogen to react with oxygen, there's no water. Maybe that doesn't rule out all life forms, but it certainly rules out everything that has ever lived on Earth.

So what happens if we decrease the strong force to, say, 0.9 on the dial? Again, once we push the red button, we get a shock. We see the flash of the Big Bang and the short, furious burn of the first stars, but they fade without going supernova. Galaxies form as before, but again their stars burn only briefly before fading into darkness. All stars have short lifetimes, big, medium and small alike, and, even worse, without material from supernovae there are no planets. We have made another universe that is barren of life. What went wrong?

The answer, again, was in the intense heat and pressure of the primordial fireball. A decrease in the strength of the strong force has meant that protons no longer fused with neutrons to make the hydrogen isotope deuterium (^{2}D), a vital stepping stone in the synthesis of ^{4}He. Without deuterium to pave the way, we effectively turned off nuclear fusion like a tap. There are no other atoms except hydrogen, no chemistry and therefore no life. Stars still light up, heated by gravitation, but without nuclear fusion reactions in their cores, they quickly lose energy as heat and cool into black holes.

Sobering, eh? We can't tweak the strong force by more than a tenth in either direction without nixing the periodic table, chemistry and life. Since the strong force competes with the electromagnetic force, that implies we can't tweak the electromagnetic force by much either without having the same effect. Why? Reduce the electromagnetic force, and ^{2}He will be stable, removing all hydrogen from the Big Bang. Increase it, and the Big Bang makes no deuterium, and hydrogen is all we get.

A PUT-UP JOB

And if none of that convinces you, try this extraordinary fact, first pointed out by Fred Hoyle: every element heavier than helium should be rare in the universe. The reason is that there's another roadblock early on in the fusion process that goes on inside stars. The easiest way to make larger and larger nuclei is to start with a helium nucleus, say, and keep adding a proton; however, there's a problem. If we add proton to a helium nucleus, we get lithium-5, which rapidly decays. If we squash two helium nuclei together, we get beryllium-8, which is also unstable. That means there's no easy way to make boron-10, whih has five protons, or carbon-12, which has six protons, or oxygen-16, which has eight. Yet carbon is the next most abundant element after hydrogen and helium, and oxygen is the next most common. In fact, apart from a scarcity in lithium, beryllium and boron, elements all the way up to iron are common-place. So what's going on?

One possible route to making carbon is for three helium nuclei to fuse together, in what is called the 'triple alpha' process. But there's a problem. The first step is for two helium nuclei to come together to make beryllium-8, which is unstable. The chance of a third helium nucleus colliding with beryllium-8 in time to make carbon-12 is therefore vanishingly slim. Hoyle's genius was to suggest that there might be a fortuitous resonance, in the form of an excited state of carbon, which exactly matched the energy of the beryllium-8 nucleus when capturing a helium nucleus. The experimentalists went to work, and, sure enough, exactly just a state was discovered.

Without Hoyle's resonance, next to no carbon would be made in stars, and – since to make an oxygen nucleus all you need is for a carbon nucleus to capture another helium nucleus – next to no oxygen. Give or take the odd rogue metal ion, all life on Earth is essentially made up of just a few elements: carbon, hydrogen, nitrogen, oxygen, phosphorus and sulphur. Can it be coincidence that these exact elements happen to be the most abundant in the universe? A change of a fraction of a per cent in the strong force, and just a few per cent in the charge on a proton would cause the Hoyle resonance to disappear. And then where would we be? Literally nowhere.

AN ACCIDENT OF HISTORY

So how do we explain this fine-tuning? Is it written into the laws of physics? So far as we know, it's not. The present laws of physics are a blank slate. They tell us the relationship between physical quantities, but not their absolute values. Newton's Law of Gravitation, for example, tells us how the gravity of an object varies with distance. Double the distance, it says, and you get a quarter of the gravity. But how much gravity does a mass of one kilogram have at a distance of one metre? Newton can't tell you. To work that out, someone somewhere has to do an experiment, make some measurements, figure out the strength of gravity per unit of mass and distance, and then plug it into Newton's equation.

And it's the same throughout physics. The Standard Model doesn't predict the masses of the fundamental particles, for example, which is why the Higgs gave us the runaround. We only know their masses from experiment. With the partial information we had, we were able to put an upper limit on the mass of the Higgs, but that was about it. Likewise, General Relativity in cosmology doesn't tell us the strength of gravity or the size of the cosmological constant. We have to build ourselves a telescope and have a look.

The truth is that over the past two centuries, experimental scientists have done an extraordinary job of keeping the theorists in business by making incredibly accurate measurements of all the constants that we find in nature. Don't think for a second that any of the theorists like this situation. The grail they chase is a theory that will, of its own sweet accord, predict the universe that we find ourselves in. This elusive Theory Of Everything (TOE), as it is known, would make all experimentalists redundant overnight. Every detail of the universe – the mass of dark matter particles, the strength of gravity, the charge on an electron – would pop out like a rabbit from a hat. Instead of using quantum physics here and relativity there, everything would be a special case of the TOE. Cosmology and particle physics would be a job well done.

Big deal, you might say. We're just not there yet. It's not surprising our theories don't describe the universe exactly as we find it; after all, we haven't got the full picture. Our 'laws' are really all special cases, like Kepler's Laws of Planetary Motion turned out to be special cases of Newton's Law of Gravitation, which are, in turn, a special case of General Relativity. Once we have a truly fundamental theory, we'll have a truly detailed description. When we do, General Relativity will prove to be a special case of some as yet undiscovered Theory Of Everything, and we'll be able to predict the rate of expansion of the universe before we measure it, not fudge it after.

But there's an alternate argument. It says that the reason we can't predict the fundamental constants is that they are a result of the universe's history rather than a feature of any fundamental theory. The universe wound up the way it is by chance, not design. Why should we be surprised that gravity is weak, when if gravity were strong we wouldn't be here?

Maybe there are other universes out there where gravity is strong; they just don't have people in. The universe is fine-tuned for life because, if it weren't, we wouldn't be here.

THE MAN WHO BROKE THE BANK

The idea that the universe is fine-tuned for life is called the anthropic principle. The fundamental constants are what they are because otherwise we wouldn't be here to ponder them. By sheer chance the strong force can overcome the electromagnetic force to create atoms, and gravity is weak, so those atoms form stars and planets. By remote accident, the atoms on those planets happen to have interesting chemistry, and the environment on at least one of those planets kick-started the very special set of chemical reactions we call Lady Gaga. We lucked out, simple as that.

Some would leave it there, content to believe that the universe is a lottery, and our numbers just happened to come up. Others – and I am one of them – would say that's not a satisfying answer. To me, a freak fine-tuned universe begs further explanation. How many other tickets were there in the lottery? How many other universes failed to create life before ours succeeded?

The philosophers among you will no doubt protest that I am falling for one of the oldest logical faux pas in the book, a version of what is known as the Monte Carlo fallacy, after a famous Monaco gambling incident. Purportedly, on 13 August 1931, the roulette wheel at the Grand Casino in Monte Carlo landed on black twenty-six times in a row. Did anyone make any money? Yes, the casino did. Because after the wheel had landed black around fifteen times in a row, everyone started putting all their money on red. Time after time they piled high their chips, and time after time the wheel came up black.

So what was the flaw in the punters' logic? Simply that the roulette wheel somehow 'knew' how many times it had come up black, so was bound to come up red. Of course, the wheel knew no such thing, which is why everyone took a very deep bath. The lesson, of course, is that truly random processes know nothing of their own history.

Take lotteries, for example. The version we have here in the UK asks you to choose six numbers between 1 and 49. After a short burst of excruciating live entertainment, the numbers are drawn live on television. The chances of matching all six numbers and winning the jackpot are 1 in 13,983,816.[18] Are the odds better if you've played a great many times before and lost? Your cheating gambling heart might say 'yes', but your cool, probabilistic head definitely says 'no'. Guinevere doesn't know how many times you've played, or what your lucky numbers are. You could just as well win at your first attempt as at your nine hundred and ninety-ninth.[19] Win the lottery one week, and you've got just as much – or as little – chance of winning it the next.

By analogy, then, am I committing the Monte Carlo fallacy when I say that a rare event like a fine-tuned universe implies lots of previous attempts which weren't fine-tuned? Isn't that just like saying, 'Hank just won the lottery. He must have been playing for years.' At the risk of answering

18 Chance of picking the right number = (chance of picking right ball) x (number of ways you can arrange six numbers) = $(6 \times 5 \times 4 \times 3 \times 2 \times 1) \times 1/(49 \times 48 \times 47 \times 46 \times 45 \times 43)$ = 1/13,983,816 or 13,983,816 to 1.

19 She doesn't know what your numbers are either. I'm not saying that having lucky numbers isn't fun, of course it is. But 1, 2, 3, 4 and 5 have just as much chance of winning as 5, 7, 24, 25 and 30 (our family birthdays). Come to think of it, I'm off to the corner shop.

my own rhetorical question, yes, I think it is the same. Just because Hank's numbers came up once doesn't mean he's played before and lost. But, then again, it doesn't rule it out either. If we raid Hank's house on a hunch, and find a cupboard full of scrunched-up lottery tickets, we wouldn't be that surprised.

And if you ask me, that's exactly what we have found with the inflation of the universe. Most, but not all, theories of inflation imply that the universe we find ourselves in is just a small part of a much bigger whole. Maybe after the Big Bang, some regions of spacetime inflated more rapidly than others, becoming isolated bubbles. Maybe what we call the universe is just inside one of those bubbles, and out there in Never-Never-Land are countless others. Our universe happens to be fine-tuned for life; maybe lots of the others aren't. It's time to meet the multiverse.

THE UNIVERSE'S STRETCH MARKS

As Carl Sagan said, extraordinary claims require extraordinary evidence, and, as far as the multiverse goes, we don't quite have the quality of proof that one might hope for. Nevertheless, we are making progress. We already had circumstantial evidence for inflation – and, by proxy, the multiverse – in the exact size of the temperature variations in the cosmic microwave background (CMB), which the theory manages to predict with impressive accuracy. The real clincher, however, would be the detection of primordial gravitational waves.

In order to understand what gravitational waves are and why they might provide evidence for inflation, I am going to have the pleasure of introducing you to one of my favourite bits of physics. In essence, it's this: acceleration makes waves. Perhaps the most obvious example is sound, where the acceleration of a solid object produces sound waves. But the same principle is at work in field theories like electromagnetism, where an accelerating electric charge radiates photons, also known as light.

Einstein's Theory of General Relativity describes how matter and energy relate to gravitation, so it fits the picture that an accelerating mass radiates gravitational energy in the form of a ripple in spacetime called a gravitational wave. That supernova on the far side of the galaxy, that pair of neutron stars orbiting one another in deep space, the tennis ball in the men's final at Wimbledon – all of them create gravitational waves. Those waves spread out through the cosmos, stretching and compressing space and time like the ripples on a pond. Eventually, they will pass through you, and your space and time will wobble. Only a bit, of course. You will shrink and stretch in height, but imperceptibly. Your watch will run fast and slow, but by such a minute amount as to be unnoticeable. And then the wave will pass, and your space and your time will be still again.

That's the theory anyway. Relativity tells us that in most cases the wobbling of space and time caused by gravitational waves will be so small as to be undetectable. To see the effects, you need to accelerate something really big. In the case of inflation, that something is the universe. During inflation, the universe expanded faster than the speed of light.[20] And when you accelerate something as big as the universe as much as that, you are going to get some pretty

---

20 I know what you're thinking: nothing can travel faster than the speed of light. Actually, that's not quite true. The more accurate statement is that no signal can travel faster than the speed of light. That leaves the expansion of space to do what it pleases.

big gravitational waves. It's the after-effects of these waves that we are currently searching for.

FLEXING OUR BICEP

Theory predicts that when the universe inflated, the gravitational waves that were produced should have left an imprint on the cosmic microwave background. To cut a long story short, they should have polarised it, leaving a distinctive pattern that a telescope in Antarctica named the Background Imaging of Cosmic Extragalactic Polarization (BICEP II) has been built to detect.[21]

In 2013 everyone got very excited when the BICEP team thought they'd cracked it, but the result has since been shown to be a false positive caused by cosmic dust. After a bit of a rethink and a chunky upgrade, they are expecting to publish more results in 2016. If they are successful, everyone will have to start taking inflation and the multiverse a lot more seriously.

There's a beautiful circularity to that. At the time of the Ancient Egyptians, we believed that mankind was special, and that the Earth stood at the centre of the cosmos. We call this the Ptolemaic Model, after the Greco-Egyptian astronomer Ptolemy,[22] who first formalised it. Copernicus then demoted us to a middle-ranking orbit around the Sun with his heliocentric Copernican Model. Einstein's General

21 BICEP II is a refracting telescope, and is based at the Amundsen–Scott Station in Antarctica. The 'z' in 'Polarization' is not a typo; the experiment is a project of Caltech.

22 Although they all hailed from Alexandria, Ptolemy the astronomer lived after, and was no relation to, Ptolemys I–XV, the last pharaohs of Ancient Egypt. Ptolemy's dates are c. AD 100–170; those of the Ptolemaic dynasty are 305–30 BC. As I say, more about the Ancient Egyptians coming up.

Relativity taught us that even the Sun wasn't that big a deal; no part of spacetime is privileged over any other. And now the entire universe turns out to be just one ticket in a colossal lottery called the multiverse.

The kicker is that makes us special after all. Or, to be more precise, our universe is special because it's the one where the very fabric of existence is just right for life. When we go looking for our cousins in the cosmos, we should bear that in mind. Because even before we came along, the universe was trying for a baby.

CHAPTER FIVE

Life

In which the author's attempt to grasp exactly what makes life special causes him to fall through a mathematical rabbit hole, only to emerge in a wonderland where everything is valued not by its beauty, but by the rate at which it dissipates energy.

On 7 August 1996, President Clinton announced to the world's media that NASA had found evidence of life on Mars. A team led by David McKay, the geologist who had trained the Apollo astronauts how to search for moon rocks, had found fossilised bacteria in a Martian meteorite. 'If this discovery is confirmed,' Clinton intoned, 'it will surely be one of the most stunning insights into our universe that science has ever uncovered.'

Sadly, it wasn't confirmed. McKay's discovery was soon submerged in a blizzard of critical academic papers. The main objection was to the fossils' size. The largest among them were only 100nm in diameter, whereas the smallest

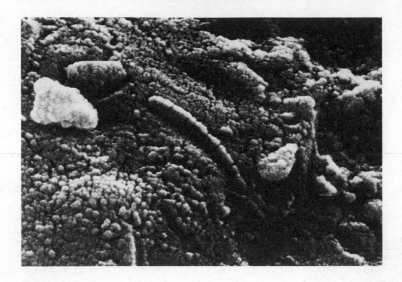

bacteria known at the time were nearly ten times that size.[1] Some critics pointed out that's too small to hold enough DNA for them to replicate. You might ask why Martian bacteria would necessarily contain DNA, but that's a whole other question.

Take a look at the photograph above, reproduced from the McKay team's paper in the journal *Nature*. I hope to convince you that the sausage-shaped 'bacteria' in meteorite ALH84001 may indeed be microscopic aliens. Even if I fail, the whole question of how we decide whether or not something is, or has ever been, alive is right at the heart of our quest to find life on other planets. If we hope to find life elsewhere, we have to know exactly what we're looking for.

1 A nanometre is a billionth of a metre, or if you like, a millionth of a millimetre. The nano scale is really the doorstep to the world of atoms, which tend to be typically around a tenth of a nanometre in size, also known as an ångström (Å). To put it mathematically, $1Å = 10^{-10}$m, $1nm = 10^{-9}$m.

BOLT FROM THE BLUE

A meteorite, of course, is simply a piece of rock which has come to Earth from space. All kinds of rubble from asteroids and comets are constantly hitting the upper atmosphere, but most of it burns up long before it reaches the ground. The result is a shooting star, or meteor. A classic example is the Perseid meteor shower, which happens every August in the northern hemisphere as the Earth trundles its way through the tail of the comet Swift–Tuttle, and very beautiful it is, too.

Occasionally, however, we get lucky – or unlucky – and a particularly large lump of space rock will touch down somewhere on our beautiful blue planet. And there's no limit, really, on how big such a rock might be. As we shall soon learn, the early solar system was buzzing with comets and asteroids, and the Earth took a severe battering from its birth right up until around 3.9 billion years ago.[2]

Things have calmed down a lot since, but now and then we still get a haircut. The last really big space rock to do the business hit us sixty-six million years ago,[3] wiping out the dinosaurs. Five thousand years ago an asteroid struck the Indian Ocean, causing a worldwide tsunami and quite possibly founding the myth of Noah's flood. And as recently as 1908, a comet exploded in the

---

2 Some of those comets and asteroids were laden with water, which is where we believe the Earth's oceans came from. They also brought enormous quantities of carbon compounds, which is where we come from. The worst of the pummelling came at the end, and is known as the Late Heavy Bombardment. We don't know exactly what caused this final flurry, but one of the more convincing explanations is that of the Grand Tack model, which claims that the giant gas planets Jupiter and Saturn hit a resonance and changed their orbits, pulling the gravitational rug out from under the toddler solar system.

3 The latest data has revised this figure; it used to be sixty-five million years ago.

skies above Tunguska, Siberia, with a thousand times the energy of the atomic bomb that was dropped on Hiroshima.

There's nothing that uncommon, in other words, about meteorites per se. What made ALH84001 so unusual is that it came from Mars. As you might have gathered, our atmosphere actually protects us from space rocks, as they tend to burn up in it due to friction. Other planets with thinner atmospheres get a rougher deal. When a stray asteroid or comet collides with Mars, the red planet takes it right on the chin. If the impact is big enough, lumps of Martian crust get billiarded out into space. If that debris escapes Mars' gravity, it can find itself on a collision course with the Earth. And in the case of our hero meteorite, the exact location of that collision was the Antarctic.

HOLES IN SNOW

The Antarctic is a meteorite hunter's heaven. One rather straightforward reason is that they show up really well against the snow. Another is that, generally speaking, ice piles up in the middle of Antarctica, and then flows down and out to its coasts, where it gets trapped at the feet of mountain ranges. Anything that falls out of the sky into the middle of the continent therefore ends up on a kind of slow-moving conveyor belt to the bottom of the nearest mountain, ready to be collected by some hardy geologist.

As a result, every summer in the southern hemisphere the USA's National Science Foundation supports the Antarctic Search for Meteorites, or ANSMET. Round about December, a small research team head out from a base camp near the foot of the Transantarctic Mountains to hunt for meteorites. The rock that the McKay team studied had been found by one Bobbie Score during a snowmobile ride on 27 December 1984, in the Allan Hills. As the first sample found that year, the rock had been labelled ALH84001.

To begin with, ALH84001 was assumed to be the remains of a common or garden asteroid, but by 1993 it was resident at the Johnson Space Center where it was identified as a piece of Mars' crust. It was at this point that it piqued David McKay's interest. Tests showed that it was an astonishing 4.5 billion years old, having formed on Mars just a squeak after the solar system itself. Sixteen million years ago some sort of impact cannoned it up and out of Mars' gravitational field,[4] where it wandered the solar system before rubbing up against the Earth's atmosphere and crash-landing in the Antarctic around 11,000 years ago.

The rock itself was riddled with cracks, having suffered some sort of impact around four billion years ago while it was still on the surface of Mars. And here was the really exciting bit. Lodged in the cracks were granules of calcium carbonate. To a geologist, calcium carbonate means water,[5] and water, as we all know, means life. Even more intriguingly, calcium carbonate granules of a similar size and shape to those in the meteorite were known to be produced by certain kinds of terrestrial bacteria.

'So what?' you may say, 'the granules probably formed on Earth during the 11,000 years that the rocks had been lying in the Antarctic snow.' But when the team dated the

4 To give you some evolutionary perspective, the great apes, or hominids, split from primates around fifteen million years ago, one million years after ALH84001 left the surface of Mars.

5 Whenever water rich in minerals absorbs carbon dioxide, calcium carbonate is never far away. In fact you probably last saw some as limescale on the heating element of your kettle.

granules, they found that they were billions of years old, not tens of thousands. There was no doubt about it; they had formed on Mars. But how? The McKay team decided to put their pet rock under an electron microscope to see if they could find out.

It was then that they saw the thousands of sausage-like shapes clinging to the granules of calcium carbonate. Have another look at that photograph, and tell me again whether you think they look like fossils of living things. If you think they do look like they were once alive, what is it about them that convinces you? And if you don't, why not?

WHAT IS LIFE?

We humans are expert at spotting life. NASA's recent Discovery mission to Mars was exhilarating, but there is something infinitely frustrating about a robotic probe scratching around for water in dusty soil when there could be purple alien bacteria crawling all over the rocks of a nearby mountain stream. Stick any microbiologist on the Red Planet for a day, and he'll be able to tell you whether there's anything living there or not. So what is it that we look for? How do we identify something as being alive?

One thing we look for is movement. Life gets around, be it a bacterium waggling its flagellum, or a flying squirrel spreading its paws and taking to the wind. As a physicist might say, living things do work, meaning they are capable of converting chemical energy into mechanical energy. The bacterium beats the surrounding liquid with its flagellum; the flying squirrel pushes down on the air to create uplift. Non-living things can do work, too, but unless they are man-made, they don't have an internal energy supply. Kick a football, and it starts to slow down the instant it leaves your foot. Kick a kangaroo and you may well end up on the other side of the nearest hedge.

And living things grow. If you're out for a Mars walk and you see a large black rock, and the next day it's twice the size, it's time to take some photos. And if the day after that the large black rock has a little black baby rock sitting next to it, you've really struck the motherlode, because the special type of growth that we call reproduction is another tell-tale sign of life. In fact, some would say it's a defining sign, because without it natural selection would have no way to work, and there would be no evolution of species.

Doing mechanical work, growing and reproducing are all things that life does, but it's interesting to note that they don't define it. An avalanche is capable of doing work, even though that work might be flattening ski chalets. Likewise, rust can grow on a chain-link fence, but we wouldn't keep it in a jar as a pet. Simon Cowell was unable to reproduce without the assistance of Lauren Silverman; nevertheless he is most definitely alive. And it's not hard to imagine some highly evolved creature of the future that decides it doesn't want to die, thanks very much, uploads itself into some sort of virtual reality and does away with natural selection altogether.[6] What is it that life does that non-life doesn't?

The thing about fossils, of course, is they don't move, grow or reproduce. If David McKay's team had been really lucky, one of ALH84001's bacteria might have been pre-served in the act of cell division, but no such luck. Neither were any of them spotted at different stages of growth, or

6 For some reason I am still picturing Simon Cowell.

flexing minuscule flagella. No, what makes the tiny sausage-shaped structures in the Allan Hills meteorite look lifelike is a truly fundamental property of all living things: they are organised.

In organising themselves, living things go against the grain of the entire universe. As we are about to see, the cosmos seeks one thing, and one thing only: equilibrium. The natural world doesn't want you to play backgammon, or learn to salsa, or even exist. It wants equal temperature, maximum disorder and death. Eventually it will get its way, but hopefully not before you've finished reading this chapter, and got to grips with one of the most mind-expanding principles in the whole of science: the Second Law of Thermodynamics.

THE RULES OF THE GAME

The concept of energy occupies hallowed ground in physics. I don't think it's too much of an exaggeration to say that it underpins every other physical theory we have, and that goes for both Quantum Mechanics and General Relativity. It is enshrined in the four Laws of Thermodynamics, which can be roughly summarised as follows:[7]

(0) If two different objects are in thermal equilibrium with a third object, then they will also be in thermal equilibrium with each other.

(1) The total energy of the universe remains constant.

---

7 I know, I know; although there are four laws, they are labelled zero to three. The problem was that the First, Second and Third Laws were already well known before it was quite rightly pointed out that the Zeroth Law – which defines temperature – was fundamental to all of them.

(2) The entropy of the universe always increases.[8]

(3) The entropy of an object approaches zero as its temperature approaches zero.

Everything that we can imagine an extraterrestrial life form might do – be it thinking, communicating, growing, moving, or reproducing – requires energy. Although there's a slightly forbidding ring to their name, the Laws of Thermodynamics are really just a very simple set of rules for the way that energy works. They are all directly relevant to life, but it's the Second Law that has the most surprising implications for living things. As we are about to see, the reason you ate breakfast was not just because you needed a source of fuel. You ate it for its information content.

DOWN WITH EQUILIBRIUM

Balance, in psychological terms at least, sounds like a wonderful thing. In fact I have often experienced it myself, if only when swinging from beetling elation to swingeing depression. In people, it is generally and justifiably admired. Who doesn't seek to be stoic, good-humoured and harmonious of spirit? The Buddha, one imagines, didn't get into bar fights, and Buddhism would be a less impressive religion if he had. But while equanimity is an enviable quality in human beings, out there in the universe at large it is very bad news indeed.

The universe, I am sorry to tell you, is not a fan of yours for one very simple reason: it doesn't like hot things

8 We'll get to entropy in a minute.

and cold things. Rather, it infinitely prefers something else: thermal equilibrium. You probably already have a rough idea of what that means, but, as a refresher, let's imagine putting something hot – a cup of hot tea, say – next to something cold, like a saucer. What happens next? No prizes for guessing that heat flows from the cup of hot tea to the saucer. Eventually, the cup and saucer will reach the temperature of their surroundings, and no more heat will flow.[9] We say then that they are in thermal equilibrium. This, of course, is what the Zeroth Law of Thermodynamics is telling us; thermal equilibrium is what results when two different objects have the same temperature.[10]

As you may know, this type of heat flow – from one body to another in direct contact – is called conduction. As far as a cup and saucer goes, the process is slow for the simple reason that the very thing they are both made out of, china, has been chosen because it is so bad at conducting heat. Our ancestors, in their wisdom, wanted their tea hot and chose their crockery accordingly. But that's not the end of the story. A hot cup of tea also cools by another method, known as convection. Basically the hot tea and teacup heat the cold air around them, and the hot air rises, only to be replaced by more cold air. Eventually the tea reaches the same temperature as the surrounding air, in which case – if you are my wife – at this point you decide to try and drink it.

Conducting and convecting heat is the kind of thing the

9 If you really want to be precise, heat doesn't stop flowing, it's just that as much of it flows from the cup to the saucer as from the saucer to the cup.

10 In case that still isn't clear, just substitute the word 'thermometer' in place of 'third object' and you'll see how the Zeroth Law is essentially a definition of temperature.

universe loves to do, because both activities take it one step closer to its ultimate goal of universal thermal equilibrium. But there's another method by which hot things lose heat to cooler things, and as far as the universe at large is concerned, it's much more important. That method is called radiation, and it's really worth understanding in detail for two reasons. Firstly, because it supports the majority of life on our planet, and, secondly, because it will ultimately bring about a dull, dull Armageddon.

RISE AND SHINE

So here goes. Every object in the universe radiates light, centred around a peak wavelength. Light, as you'll remember,[11] can have a whole spectrum of wavelengths, starting with radio waves, then on to microwaves, through the infra-red to visible light, and on up through the ultra-violet to x-rays and finally gamma-rays. The higher an object's surface temperature, the shorter the peak wavelength of the light it emits. In case that sounds a bit dry, let's talk about a concrete example. You.

The core body temperature of the average healthy human being is, as most people know, roughly 37°C. It won't surprise you to learn that your surface temperature when clothed is somewhat less than that. Assuming you are relaxing in a reasonably well-heated room at around 20°C, perhaps wearing a onesie, you might expect it to be something in the region of 28°C. Any object – and I mean *any* object, it could be a piece of granite, a plastic statue of Elvis, or a possum – with a surface temperature of 28°C

11 If you've been diligent enough to read the footnotes.

will radiate light, and the peak wavelength of that light will be roughly ten millionths of a metre, in the region of the spectrum we call the infra-red.

Maybe you've worn a pair of thermal imaging goggles, or, as in my case, seen Walter White doing so in an episode of *Breaking Bad*. Either way, you will know that people show up bright, and almost everything else shows up dark. That's because the goggles are capable of detecting infra-red light and using it to form a visible image. The shorter the wavelength of the infra-red, the brighter the image in the glasses. When Walter looks at Jesse, Jesse will show up brighter than his surroundings because he is at a higher temperature than they are. Should Jesse be holding a cup of hot tea, then that would appear brighter still.

HERE COMES THE SUN

So every object in the universe gives off light, and the hotter it is the shorter the wavelength of the light it gives off. The very hottest things in the universe, of course, are stars. Our own Sun has a core temperature of some 16 million °C, and a surface temperature of 5.5 thousand °C.[12] The light it emits spans the entire spectrum from radio waves to gamma-rays, but it peaks around the infra-red to visible part of the spectrum.

Colour, in other words, is closely linked to temperature. Your gas ring burns blue, and the embers in the fire burn

---

12 Coincidentally, the surface temperature of the Sun is roughly the same as the core temperature of the Earth. The much-more-trustworthy-than-Wikipedia maths and science website Wolfram Alpha quotes 5,777K for the Sun, and 5,650K for the Earth. 1 Kelvin = – 272.15°C.

red, because gas burns at a higher temperature than wood. Temperature, and temperature alone, dictates the colour of the light an object radiates. Heat a banana to 5,500°C and it will give off the same characteristic blend of colours as the Sun.[13]

So what happens to the light radiated by the Sun, after it strikes the Earth? The short answer is that it bounces around like nobody's business until it finally gets absorbed. That Ferrari on your drive is red because the paint is absorbing every light photon that hits it *except* those that are red. These reflected red photons will eventually land on something that will absorb them; your monogrammed black leather driving gloves, perhaps.

After they have absorbed the red photons, and any other photons that happen to strike them, the temperature of your driving gloves will increase. They will then radiate a spectrum of electromagnetic energy, peaking in the infra-red, which will in turn be reflected or absorbed by any other object – the obsidian-grey leather steering wheel, the tan calf-leather seats, your pink jumpsuit – that they come into contact with.

I'm sure you can guess where this is heading. In this universe of ours, objects don't have to be in contact with one another for heat to spread between them until they reach the same temperature. It can happen just as easily in a vacuum, via the process of radiation. In fact, because most of the universe *is* a vacuum, radiation is by far the most common

---

13 Viewed from space, the Sun is white, because it radiates across the entire visible spectrum, with no wavelength being dominant enough to affect the overall colour. Here on the ground, however, direct sunlight is most definitely yellow. That's because whereas red light travels unhindered through the atmosphere, blue light is easily deflected by air molecules. Physicists call this scattering, and it means the atmosphere effectively 'holds on to' the blue portion of sunlight, making the Sun yellow and the sky blue.

way that heat gets around. Seen this way, the reason the stars shine is indeed to guide your destiny. It's just that your destiny happens to be thermal equilibrium with the rest of the universe.

So what has this got to do with life? Everything, it turns out. Life is an extremely perverse cooling process. The flow of heat in the universe may be a one-way street, from teacups to saucers, planetary cores to atmospheres, and from stars out into the cosmos, but it can be harnessed by life forms to do work, reproduce, and – most importantly – organise themselves. It's time to meet the First Law of Thermodynamics.

PINBALL WIZARD

The First Law is great, because it effectively provides an accountancy system for the entire universe. Let's remind ourselves of what it tells us:

(1) The total energy of the universe remains constant.

Put simply, this is nothing more than conservation of energy. Fire a pinball machine, and you may watch impotently as the ball careers around, missing every jackpot and reward channel before it finally rolls with unswerving accuracy directly between the flippers and into the drain. But the beauty of energy is this: if we know the energy of the fired ball, and the energy of the returning ball, by subtraction we can work out exactly how much energy has been absorbed by the machine, despite knowing nothing at all about what went on in the game.

How is this relevant to our search for life, you ask? Well,

like our pinball machine, cells consume energy, and expend that energy as work and heat. To understand politics we follow the money; to understand cell biology we follow the energy. Take any kind of cell, and we can ask some simple but far-reaching questions. How does it get its energy? How does it store that energy? And what does it use it for? As we shall see in the next chapter, when we ask these questions of the very first bacteria we get fascinating clues as to how life on Earth got its start.

But back to the plot. The First Law tells us that energy may be converted from one form to another, but it can't be created or destroyed. The pinball machine is a classic example. When we pull back the pin, we convert chemical energy in our muscles into the mechanical energy of a compressed spring. Release the pin and the spring's mechanical energy is converted to kinetic energy as the ball flies away at speed. The tilt of the machine then converts some of the ball's kinetic energy into gravitational potential energy as it rolls up the slope of the machine, rising in the Earth's gravitational field. And that's before the game has even properly started.

In a perfect world, at every one of these conversion stages no energy would be lost. Each type of energy would be completely converted into the next, and, theoretically at least, a turn at pinball could go on forever. But real life, as we all know, is not like that. Compressing a spring, rolling a metal ball across a wooden surface and just about everything else all generate heat. What's worse, that heat leaks out into the universe and gets absorbed. You can never get that energy to do anything useful again; it is lost to you forever. The quest to understand why – despite energy being conserved – you never get as much out of a system as you put in is what led to the extraordinary bit of physics that is the Second Law. Because it turns out that one of the reasons that the globules in the Martian meteorite look like they might have once been alive is that they appear to break the Second Law of Thermodynamics.

ONLY HERE FOR THE BEER

Even if you don't know what the Second Law is, you will almost certainly have heard of it. Thomas Pynchon, one of my favourite American authors, is obsessed with it, and if you are an enthusiast of abstruse fiction I thoroughly recommend his novel *The Crying of Lot 49* for its deep insights into what has to be the most fascinating of all physical laws. To really understand it, we first need a deeper understanding of something we have so far taken for granted: heat.

Our goal, as ever, is a deeper understanding of biological systems, but the work and character of a nineteenth-century brewer named James Prescott Joule are so especially noteworthy I can't resist a momentary diversion. The brewing game requires precise measurement of temperature, and Joule carried the art to virtuosic levels. He became fascinated with exactly what it was he was measuring, and in a landmark experiment demonstrated that what had previously been thought of as a fluid was in fact the random motion of atoms.

Put simply, heat is atoms jiggling about. The hotter something gets, the more its atoms jiggle. Put something cold – or, in other words, unjiggly – in contact with something hot, and the jiggly hot atoms will eventually jiggle the unjiggly cold atoms into some sort of intermediate jiggly-ness. Thermometers measure this jiggly-ness. This, of course, is why temperature has a zero, because in theory it's possible for something to have no heat and therefore not jiggle at all.[14]

Joule's trick was to drop a precisely known weight through a precisely known distance, making it do work on the way down by spinning a paddle in a bath of water. The spinning paddle pushes the water molecules around, increasing their speed and therefore their temperature. By accurately measuring the temperature increase in the water, Joule was able to demonstrate that the increase in its thermal energy exactly equalled the gravitational potential energy imparted by the falling weight. At a stroke, he showed that heat is basically a type of atomic motion. Even today, it still seems somehow revolutionary. But there you have it. Sloshing water about increases its temperature.[15]

THE INFORMATION ENGINE

Right. We're ready for the Second Law. At first sight it appears to be one of the most unhelpful bits of science you could wish for, a non-event of the first water. Do not be fooled. You are about to go down a rabbit hole that will lead you deep into the innermost workings of the universe. Here she is:

14 Well, almost. Quantum Mechanics shows us that even at absolute zero, a particle has a residual jiggly-ness, known as zero point energy.

15 The unit of energy, the joule (J) is named in his honour. The one used by physicists, anyway. Engineers use handier (i.e. much, much larger) units like the kWh, where 1 kWh = 3 600 000 J.

(2) The entropy of the universe always increases.

So what exactly is entropy? Step by step, our understanding has deepened. It was first defined during the age of steam. A steam engine, as you will know, works by burning fuel to boil a body of water, and then using the steam created to drive its pistons. The steam then liquefies in a cold condenser before being reheated. The problem that quite rightly intrigued capitalists of the age was how to arrange this system of hot tank (boiling water) and cold tank (condenser) in order to generate the most mechanical work for a given amount of fuel.

The problem tested some of the finest scientific minds of the time. First out of the blocks was the French physicist Sadi Carnot, who in true Gallic fashion published one of the most whimsically entitled works in the history of the physical sciences with his 1824 memoir, *Reflections on the Motive Power of Fire*. Carnot rightly determined that what really counted in a steam engine's efficiency was not the amount of fuel burned, but the temperature of the hot and cold tanks. But what was the formula?

If it was Carnot who framed the problem, it was the German physicist Rudolf Clausius who enabled its solution. He did so by recognising that it is not the total amount of heat energy sloshing around in a steam engine that counts, but the portion of that energy that is available to do work. That led him to consider what had happened to the portion of an engine's heat that isn't available for work. In the case of a real steam engine, for example, there will be energy lost to its surroundings, waste energy generated by friction in its pistons, and losses due to the viscosities of both steam and water. None of that energy is ever coming back, or at least it is extremely unlikely to. How to characterise these unknown, unpredictable losses?

Clausius, ingeniously, had a way. He invented a measure of this unavailable energy, and called it entropy. Remember the pinball machine? In that case, the gain in entropy of the machine and its surroundings will be that fraction of the ball's initial energy that has been absorbed by friction and heating effects during the course of a turn. Entropy, as defined by Clausius, is simply a measure of how much of the input energy had become unavailable for work.

I know, I know; it's hard to see what any of this has to do with the origin of life. Your righteous indignation is understandable. No one, frankly, was expecting the Second Law of Thermodynamics to lead where it did. Clausius just wanted a working theory for steam engines. Instead, he ended up discovering nothing less than a brand new physical quantity, an entity that appears to be as fundamental to the way the universe works as energy or temperature. Because when the time came to try and understand entropy on the scale of atoms, an astonishing discovery was made. Entropy is directly related to information.

THE DEVIL IN THE DETAIL

The first man to put this discovery on a mathematical footing was the Austrian physicist Ludwig Boltzmann. His is a sad story. A brilliant man, he suffered vicious attacks from his German contemporaries, who were both unable to follow his deliciously nimble calculations and refused to subscribe to the atomic theory that underpinned them. Prone to depression, he committed suicide in 1906, aged just forty-two, but not before crafting the equation that appears on his tombstone:

$$S = k \log W$$

In this masterpiece, Boltzmann relates the entropy, S, of a gas[16] to the number of ways its particles can be configured, W, while still having the same overall pressure, volume and temperature. The constant k, by the way, is respectfully known as Boltzmann's constant. All we're aiming for here is the gist, so the point to grasp is this: the greater the number of configurations a system has, the more uncertain we are about which one it is actually in.

How so? Let's take a concrete example. Say I take a party balloon and blow it up. I can easily measure the volume, pressure and temperature of the air in the balloon. In how many ways could the air molecules inside the balloon be configured[17] to create the precise values of volume, pressure and temperature that I measure? In W ways, that's how many, and W is a big number. The entropy of the air inside my balloon is then k log W.

Now say I burst the balloon. The air molecules that were inside now start to mix with the air molecules in the room. Given enough time, they will escape the room and mix their way around the globe. In how many ways might they be configured then? I have no idea, but I can tell you that it's a lot more than W, the number of ways they could be configured within the balloon. My uncertainty about

16 To be precise, Boltzmann's formula is that for an ideal gas. This is a gas whose individual particles don't interact with one another, so troublesome things like viscosity can be safely ignored.

17 By configured, of course, I mean where are they, what direction are they going in and how fast?

the configuration of the air molecules in the balloon has increased, and their entropy has therefore gone up.

INFORMATION IS THE RESOLUTION OF UNCERTAINTY

Following Boltzmann, the union between entropy and information was solemnised by the work of a brilliant American electrical engineer and mathematician named Claude Shannon. Although he is arguably the founder of the digital age, Shannon is criminally underacknowledged. At the tender age of twenty-one he used his Master's thesis to invent digital circuit design, the foundation of all modern electronics, and following the Second World War he was employed at Bell Labs on a US military contract.

Radio communication had become essential to warfare, and Shannon was tasked both with improving the existing military systems and making them more secure. One of the great problems in electronic devices is how to reduce the effect of 'noise' – random fluctuations in the signal. To help solve the problem, Shannon defined a new quantity, H, called the Shannon entropy, as a measure of the receiver's uncertainty as to the precise letters of a message. To be precise, he said that the Shannon entropy, measured in bits, was given by the expression

$$H = - \Sigma\, p_i \log_2 p_i$$

where H is the Shannon entropy, measured in bits,[18] of a message that is conveyed by i letters each with probability

18 No doubt you recognise this unit, the bit, as fundamental to computing. Bytes and bits are the pounds and pence of information, where 1 byte = 8 bits. When we ask for a 15 terabyte hard drive, we politely request the capacity to store a message with a Shannon entropy of $15 \times 10^{12} \times 8$ bits.

p_i and that funny squiggle is sigma, meaning 'sum up the following for all values of i'.

In case that's a little abstract, let's make these equations flesh. You and I arrange to go apple-scrumping after lights out. At nine o'clock, when it's dark – it is autumn after all – I will creep into your garden and look up at your bedroom window. If your torch is on, you are going to shin down the drainpipe and join me in Farmer Benson's orchard. If it's off, well, we are going to have to go tomorrow night instead.

In Shannon's terms, the message – 'Yes, I am scrumping' or 'No, I am not' – is coded in two 'letters', 'torch on' and 'torch off'. At one minute to nine, while I watch my breath condense in the rhododendron bushes, I have no idea whether you are coming or not. If both outcomes are equally likely, then Shannon's equation tells us that the Shannon entropy is as follows:

$$H = - \Sigma \, p_i \, \log_2 p_i$$

Two possible outcomes, each of which is equally likely, means $p_1 = p_2 = \frac{1}{2}$, giving us

$$H = - \{ \, \tfrac{1}{2} \log_2 (1/2) + \tfrac{1}{2} \log_2 (1/2) \, \}$$

Realising that $\frac{1}{2}$ is a common factor, and remembering that $\log(A/B) = \log A - \log B$ we get

$$H = - \tfrac{1}{2} \{ \log_2 1 - \log_2 2 + \log_2 1 - \log_2 2 \}$$

$$H = - \tfrac{1}{2} \{ 2\log_2 1 - 2\log_2 2 \}$$

Recalling that $\log_{anything} 1 = 0$ and $\log_2 2 = 1$, it all shakes down so that

$$H = -\tfrac{1}{2}\{-2\} = 1 \text{ bit}$$

OK. This is heading somewhere, I promise. Let's now imagine that we decide to include a third 'letter', where you waggle your torch about, meaning 'wait because I am still deciding'. Let's further imagine that each of the three possible letters is equally likely. What's the Shannon entropy then? Clearly it's more than in the previous case, but by precisely how much? Let's plug in the numbers to give us the answer.

$$H = -\Sigma\, p_i \log_2 p_i$$

$$H = -\{\, 1/3 \log_2 (1/3) + 1/3\log_2 (1/3) + 1/3 \log_2 (1/3) \,\}$$

$$H = -1/3\,\{ -3 \log_2 3 \}$$

$$H = \log_2 3 = 1.58 \text{ bits}$$

Maybe you're starting to see the pattern? For two letters, we had $H = \log_2 2$ bits. For three, $H = \log_2 3$ bits. And if we had W letters, each of which were equally likely, Shannon's formula tells us that the information entropy of the message in bits is given by

$$H = \log_2 W$$

Or changing the base of the logarithm to the natural number, e, we get[19]

---

19 In case that means nothing to you, then e is a number that crops up so often in maths it is called the 'natural number'. It's irrational, meaning that it can't be expressed precisely, and is roughly 2.718. Here's just one of the ways e can be written: $e = (1 + 1/n)^n$ as n tends to infinity.

$$H = K \log W$$

Where K is a constant. Wait a minute! That's uncannily reminiscent of Boltzmann's formula for the entropy of a group of particles with W possible configurations, each of which is equally likely:

$$S = k \log W$$

So what's going on?

SOD'S LAW

On the face of it, the progression seems unlikely. At one end, nineteenth-century physicists were looking for a way to improve the efficiencies of steam engines. At the other end, twentieth-century engineers were looking for a way to improve the efficiency of communication devices. Extraordinarily, they both turned out to be working on the same problem. Entropy, S, being our uncertainty about the microscopic configuration of a physical system, and entropy, H, being our uncertainty about the configuration of the letters in a message, are connected.

The link is information. When we calculate the entropy of a group of atoms, we can think of it either as our uncertainty as to which one of its possible W configurations it is actually in, or as how uncertain we are as to the content of a message that completely describes that configuration. Boltzmann entropy is, in fact, a special case of Shannon entropy.

So what does this mean? Well, for a start, it means that entropy is the enemy of information. The greater the entropy of a system, the less we know about the content of any message it might contain. But on a deeper level it means that information is more than just books, DVDs and hard drives; it is a fundamental property of matter. A sugar molecule, a photon of light, an edition of *The Times*: all of them contain information. That information may be to do with the location of individual atoms, or the location of George Clooney's wedding; it's all the same to the universe. It simply doesn't care for it.

Now we start to understand the true nature of the Second Law. Energy dissipates. The entropy of the universe always increases, corrupting information, increasing disorder and dispersing heat. Now we see why those fossilised grains in the Allan Hills meteorite are so striking, and, indeed, why all life forms are so magically unusual; they contain an extremely high degree of order, far greater than could arrive by chance. From stromatolite colonies of cyanobacteria, to Venus flytraps, ants' nests and bridge clubs, life forms constantly cheat the Second Law of Thermodynamics. From a world bent on chaos they concentrate energy, order and information.

Life is shit, and then you die. Everyone is promoted to a position of incompetence. If something can go wrong, it will go wrong. These are all statements of the Second Law, and they are so ingrained in us that they feel like second nature. The joy that we feel when we tidy up the office, write a letter or hold a newborn baby is a vaulting ecstasy at momentarily defying the Second Law of Thermodynamics. We know that it can't last, but somehow that makes it all the more sweet. We got one past the goalie. We passed the flaming torch of information, despite the best efforts of the universe. Life goes on.

We know, of course, that in the long run it's probably not

good news. Life is swimming against the tide of creation, and it can't last forever. In a moment, I'm going to let the Laws of Thermodynamics run to their inevitable conclusion, but first we need to answer the all-important question: how does life do it? If the universe is painstakingly eradicating information wherever it can find it, how did life manage to get started, and how has it managed to become ever more complex?

INSANE IN THE MEMBRANE

For once I'm going to give a straight answer to a straight question: cells. The fundamental building block of all life on Earth is an ingenious way of piggybacking the Laws of Thermodynamics. All life on Earth is made up of cells, and they all have one thing in common. They have a means of keeping their insides separate from their outsides.

On the face of it, a cell membrane might not be the sexiest of features. There are other, far more photogenic things to get excited about like nuclei, Golgi bodies and mitochondria, but arguably none of them would ever manage to scratch a living without something to protect them from the big bad world. As Claude Shannon might see it, a cell membrane is a great way of separating a low-entropy, high-information region – the cell's insides – from its high-entropy, low-information outsides, aka the universe. And it's this separation that enables the cell as a whole to perform a neat thermodynamic trick.

It works like this. It doesn't matter if there's a decrease of entropy inside the cell, so long as outside the cell it increases by an even greater amount. Overall, the entropy of the universe will have increased and the Second Law remains unbroken. All you need is a cell membrane to act as a gatekeeper, letting in low-entropy stuff and letting out high-entropy stuff. The low-entropy stuff we call food. The high-entropy stuff we call waste.

In other words, the cell membrane is crucial because it prevents equilibrium. Within it, entropy can be lowered, matter can be organised and information can be stored. But it also has another crucial function that was essential to the first single-celled life forms. It is a great way of storing energy.

MY NAME IS LUCA

In the next chapter we'll look in detail at what we know about the history of life on Earth. Our goal will be to try and understand how single-celled life got its start, and the series of innovations and coincidences that led to our own species of technologically accomplished and highly social apes. Once we have some kind of perspective on how likely our own intelligence is, hopefully we can get a feel for how commonplace our kind of intelligence might be in the galaxy, and how far away our nearest neighbours might be.

I'm not one for spoilers, but in the broadest of strokes we will find that the very first life on our planet was single-celled, and came in two types known as archaea and bacteria. Both originated in water. We can tell from analysing the DNA and proteins within them that they are related, but so far we have no way of telling which came first. For the moment, we are just trying to understand the nuts and bolts of how life works, so we can dispense with the gory details. What's important for our present purposes is that they both store energy by pumping protons across their cell membranes.

A proton, you'll remember, is nothing more than a lone hydrogen nucleus. Whereas an electron carries a single unit of negative charge, a proton carries a single unit of positive charge. Archaea and bacteria are capable of getting their energy from a bewildering number of sources: from eating one another, from reactions with chemicals like hydrogen sulphide and ammonia, from rusting metals such as iron, and, of course, directly from sunlight. In every case, once acquired, they use that energy to drive protons across their cell membranes, storing it up for future use.[20]

Because protons carry charge, they essentially want to get away from one another. By creating an excess of protons in the water surrounding them, and a deficit within, these simple cells are effectively creating an energy source to be tapped at will. This is arguably a bit too much detail, but for some reason I can't resist telling you that, when the time is right, these energetic protons are used to make an energy-rich molecule called adenosine triphosphate, affectionately known as ATP.

This plucky molecule is essentially the currency of energy within all cells, able to donate energy wherever it is needed. All the processes that you can think of such as movement, making proteins, RNA and DNA all extract their energy from ATP. Put bluntly, cells are miniature machines, capable of extracting energy and information from the environment, storing it, then making copies of themselves.

20 The chemist in me can't help but point out that there is no such thing, really, as a lone proton in water. Protons react with water molecules to form hydronium ions, H_3O^+. The net result is the same, though, as hydronium ions repel one another, and will readily give a proton up so that it can pass back through the cell membrane.

Single-celled organisms such as bacteria reproduce by dividing; their human cousins by having dinner then progressing to a second date.[21] Eventually, however, entropy will have its way. It's time to glimpse the end of days.

THE HEAT DEATH OF THE UNIVERSE

Let's begin by getting our bearings. The universe began from a microscopic, hot, dense state some 13.7 billion years ago, inflated in a so-called Big Bang for a fraction of a second, then settled down to a steady expansion which ended about seven billion years ago, when it was roughly half the age that it is now. At that point, some repulsive force that we call dark energy began to dominate over gravity, and its expansion began to speed up. Our best prediction is that this expansion will continue to accelerate, pushing more and more of the cosmos out of the reach of our most powerful telescopes. In fact, some two trillion years hence, the only galaxy we will be able to see will be our own.[22]

The present era is known as the stelliferous era, meaning quite simply the one where star formation takes place from the gravitational collapse of gas and dust. Before it came the primordial era, when the intense energy of the Big Bang cooled to form fundamental particles, then nuclei, then hydrogen and helium. The primordial era lasted about a

21 Dying is an innovation of sexually reproducing complex organisms, with a division of labour between mortal 'body' cells and immortal 'reproductive' cells. Bacteria, for example, are entirely immortal, though it's not a life I feel particularly envious of. There's no escape though: eventually the food will run out, and even bacteria will starve.

22 It will be a pretty big galaxy, though. The Milky Way is expected to collide with the Andromeda Galaxy in about four billion years, and in four hundred and fifty billion years' time, the fifty-odd galaxies of the Local Group will have merged.

million years; the stelliferous era will last a few trillion. So what happens next?

Well, for a kick-off, there'll be no new stars. All the hydrogen and helium gas will have been used up, and one by one the existing stars will start to burn out. The longest lived will be the smallest, the so-called red dwarves, but after a few trillion years even they will have exhausted their supply of nuclear fuel and will be beginning to cool. This is known as the degenerate era, after the extremely dense form of matter that remains when small stars cool.

There's no escaping this fate for any star, least of all our own Sun. As you probably know, our home star will run out of hydrogen in around five billion years' time, at which point it will bloat from a yellow dwarf into a red giant.[23] In roughly 7.9 billion years' time it will explode in a planetary supernova, and all that will be left is a small lump of hot degenerate carbon known as a white dwarf. In roughly a quadrillion years – that's a thousand trillion – it will have cooled to a temperature of just a few degrees above absolute zero. At that point it will no longer radiate any kind of light – radio, infra-red or otherwise – and will become what is called a 'black dwarf'.

It gets worse. As the degenerate era progresses, the swirling mass of dead stars that form the galaxy will slowly dissipate. One by one, near-misses and collisions will fling planets, black dwarves and neutron stars out into intergalactic space. Any dead stars or rogue planets that remain will slowly be consumed by black holes. At the end of the degenerate era, estimated to arrive in some 10^{43} years' time, all that will be left of the cosmos will be a silent colony of black holes, gorged to the eyeballs on dead stars.

---

23 It will reach its maximum size in around 7.9 billion years' time, at which point its surface will extend as far as the present orbit of Mercury.

That marks the beginning of what is prosaically known as the black hole era. But it doesn't end there. Black holes, it turns out, aren't completely black. As Stephen Hawking was the first to point out, subatomic particles are able to 'tunnel' across the event horizon, and return to the everyday universe. This, like any kind of radiation, has the effect of removing energy from the black hole, giving it a finite lifetime. In the case of a supermassive black hole that's something in the region of – here comes the biggest number in this book – 10^{106} years.[24] All that will survive of them will be a thin gruel of subatomic particles.

And that's pretty much all that will be left of everything else, too. Even things like protons and neutrons are expected to decay back into subatomic particles – and all so-called solid objects like planets, asteroids and comets are subject to the same quantum mechanical tunnelling that black holes are, and will eventually deplete away into nothing. Everything in creation will be a cold, cold gas of subatomic particles, jiggling away at just a few fractions of a degree above absolute zero. No part of this gas will be hotter than any other, and no life of any conceivable kind will be able to exist.

A ROCK AND A HARD PLACE

Every year my extended family makes a pilgrimage to Robin Hood's Bay on the North Yorkshire coast. One of the main attractions – apart from ice cream and fish and chips – is the fossil hunting. Edged with clay cliffs, the

24 That's calculated for a black hole of twenty trillion solar masses. The largest black hole currently known is NGC 4889 which weighs in at twenty-one billion solar masses.

pebbled beach in the bay is a treasure trove of ammonites, devil's toenails and sharks' teeth. The process is always the same. For the first twenty minutes or so you find nothing. Then you find a single fossil – usually not a very good one. Then suddenly there are fossils everywhere. Virtually every stone you pick up contains some half-submerged prehistoric creature, preserved in exquisite detail.

Seen from the perspective of entropy, there isn't a great deal of difference, really, between fossils and holiday snaps. If the Second Law tells us anything, it's that meaning has no permanence. That coiled black ammonite may feel ancient, but the whole history of life on Earth is fleeting, like a drop of spray flung high into the air by a crashing waterfall. We life forms are a glorious curiosity, another way for a star to cool, and for the universe to oxidise carbon. We are a means to an end.

If that sounds bleak, it's really not meant to. If it stirs anything in you, hopefully it's a raging thirst for what the universe wants to deny you: knowledge. Or, rather, you are a manifestation of the universe's wildest wish, namely, to awake and know itself. We have learned something profound about life. Wherever we find it, and whatever its building blocks, it will require a constant source of energy. It will use that energy to organise itself, at the expense of the entropy of its surroundings. And it will be far from equilibrium, because equilibrium means death.

CHAPTER SIX

Humans

In which the author stakes his footling reputation on one particular hypothesis of how life got its start, and forces himself through the evolutionary bottlenecks that impede the flow from micro-organism to Microsoft.

The booking hall at Euston Station will never be the same again. I am crossing it now, as I hurry to catch my train, my head buzzing with ideas. Not my ideas, I hasten to add. They are the ideas of Nick Lane, the Provost's Venture Research Fellow at University College London. Though if you ask Lane – and I just have – he will say some are the ideas of one Mike Russell, now of NASA's Jet Propulsion Laboratory in Pasadena, who first proposed them some twenty years ago.

I came to see Lane because I am on a quest. To try and figure out how likely intelligent aliens are, I need to know how likely I am. We've seen that, from its very beginning, the universe was 'trying for a baby', in that nature is fine-tuned to produce atoms. But it's one thing to try and another to conceive. How likely was it that a small, wet, rocky planet orbiting a humdrum star in a spiral galaxy would be the birthplace of single-celled life? And how and why did that single-celled life evolve into a technologically advanced civilisation of intelligent apes?

I'LL HAVE THE PRIMORDIAL SOUP

The classic picture of what scientists call abiogenesis is most often attributed to Charles Darwin. Though he avoided the subject in his public work, in a letter of 1871 to his close friend the botanist Joseph Hooker he made his true feelings clear:[1]

> It is often said that all the conditions for the first production of a living organism are present, which could ever have been present. But if (and Oh! what a big if!) we could conceive in some warm little pond, with all sorts of ammonia and phosphoric salts, light, heat, electricity, etc., present, that a protein compound was chemically formed ready to undergo still more complex changes, at the present day such matter would be instantly devoured or absorbed, which would not have been the case before living creatures were formed.

Leaving aside the shocking punctuation, what Darwin seems to be saying is that, given the right conditions, something living can spontaneously emerge from something non-living simply by chance. In essence, that has been the

---

1 Darwin was such a secretive soul that his letters to Joseph Hooker form the backbone of all Darwin scholarship, and you can find a link to them at http://www.darwinproject.ac.uk/darwin-hooker-letters

non-religious view ever since Aristotle, and it remains our belief today, though there has been a great deal of toing and froing over exactly what those conditions might be.

In the 1920s, the Russian biologist Alexander Oparin and the British polymath J. B. S. Haldane independently refined Darwin's conjecture into what is known as the 'primordial soup' theory. The gist is that once you have the right chemical elements and a source of energy, sooner or later a self-replicating molecule will emerge which is then capable of undergoing natural selection, in turn producing life.

So what might those chemical elements be? All earthly life, as you probably know, is based on one most extraordinary element: carbon. Carbon is a party animal, eager to bond with a variety of other elements – and also with itself – to form long-chain molecules and rings of almost infinite variety. In organisms, we generally find it in the company of five 'usual suspects': hydrogen, nitrogen, oxygen, phosphorus and sulphur.[2]

Two configurations can rightly be thought of as the 'building blocks' of life. One is called a nucleobase, which makes up both RNA and DNA, and also the very important energy-carrying molecule called ATP. The other is called an amino acid, the building block of proteins. I say building block; neither nucleobases nor amino acids are particularly simple, a point I have tried to emphasise by means of my characteristically poor drawings over the page.

Right. So the classic picture of the primordial soup theory in action looks something like this. It's a foul night, and a bolt of lightning strikes in the skies above

2 Often given the mnemonic CHNOPS. That's if you can call a string of letters a mnemonic.

Amino Acids:

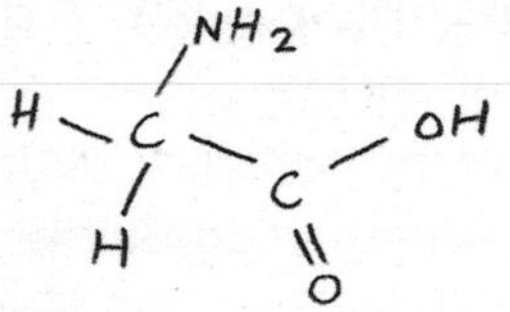

glycine

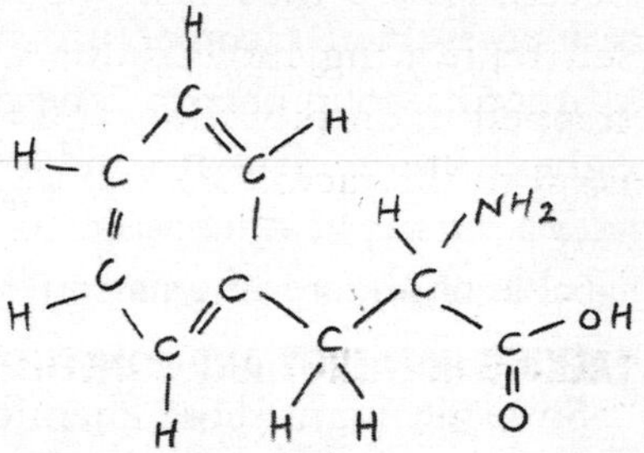

phenylalanine

Nucleobases:

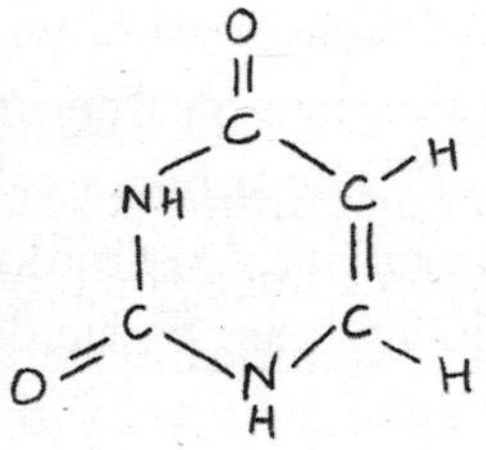

uracil

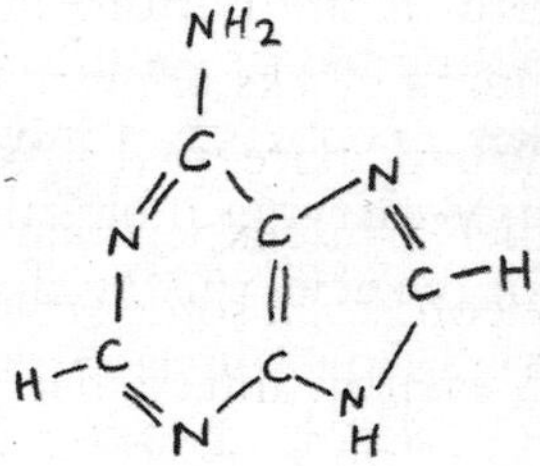

adenine

a broiling sea, in an atmosphere pumped full of noxious gases. Emboldened by this spark of energy, the gases react to form building blocks like nucleobases and amino acids. These building blocks then rain down into the ocean, where they fuse together to make the very first self-replicating molecules. A little packet of them gets trapped in an oily film, and the first cell is formed. Life is off to the races.[3]

LACKING IN ENERGY AND CONCENTRATION

Back in 1953, a Nobel Prize-winning chemist at the University of Chicago, Harold Urey, decided to put the primordial soup theory to the test. In a flask, Urey's PhD student Stanley Miller put the gases they believed to have been knocking around on the early Earth – ammonia, methane, hydrogen and water vapour – and passed an electric current through them. The results were extraordinary. There in the bottom of the flask were amino acids, the building blocks of proteins.

What's more, in the years since what is now known as the Miller–Urey experiment, we have found amino acids and nucleobases in some far-flung places: on comets, meteorites and even in interstellar space.[4] That's significant, because

3 I feel like I should point out that there are tribalisms within the primordial soup theory. One favours self-replicating molecules first; one proteins; and one cell membranes. Sooner or later, of course, you are going to need all three.

4 To qualify that sweeping statement: to date, samples of amino acids have been found in Antarctic meteorites, and their spectra have been observed in comets and interstellar space. Nucleobases have also been found in Antarctic meteorites, and the spectrum of methanamide, their chemical precursor, has been identified in comets. The spectra of amino acids have been seen in interstellar space, but not – to my knowledge, at least – that of nucleobases or their precursors.

for its first billion or so years the Earth was pummelled by comets and meteorites, with a particular flurry of blows taking place roughly 3.9 billion years ago during what is known as the Late Heavy Bombardment. The primordial seas could very well have been a soup of amino acids and nucleobases, some earthly and some alien. Could a lightning strike have kick-started life as we know it?

In the minds of many contemporary evolutionary biologists – and this sketch comedian – there are two reasons why the answer to this question is a polite 'not really'. The first is what is called the concentration problem. For a series of chance encounters to produce something as complex as a self-replicating molecule you need a lot of the right kind of molecules in the right place, which is why chemists tend to do their experiments in test tubes rather than oceans. A broiling primordial sea is simply not the place you can expect to find delicate organic chemistry.

As a workaround, some evolutionary biologists have suggested that Darwin had it right, and the vital reactions took place in a pond, where – thanks to evaporation – the 'soup' was much thicker. Which sounds promising, until you consider the fact that, as already mentioned, the early Earth was being bombarded by comets. The exact time when life began is subject to intense debate, but, to be brief, we have fossil evidence of so-called 'microbial mats' at around 3.5 billion years ago,[5] and chemical evidence of cellular life from 3.7 billion years ago.[6] As a result, most pundits would be happy to set a date for 'prototype' life at around

5 Microbial mat fossils 3.5 billion years old have been found at the Dresser Formation in the Pilbara region of Western Australia. Microbial mats are communities of bacteria.

6 The Isua Greenstone Belt in south-western Greenland, dated at 3.7 billion years old, has carbon in it that seems to be of organic origin.

four billion years ago. The really peculiar thing about that is it predates the Late Heavy Bombardment. Whatever this 'prototype' life was, it appears to have survived a cataclysm.

But for me all that fades into insignificance when it comes to the second problem: energy. As we learned in the last chapter, to sustain life we need a ready supply of information-rich energy. The 'wham, bam, thank you, ma'am' of lightning just won't do the trick; life needs a lover with a slow hand. And when I say slow, I mean *slow*; for natural selection to do its thing, we need an energy supply that lasts for tens, hundreds, maybe even thousands of years. Thankfully, in a flash of inspiration over two decades ago, the aforementioned Mike Russell pictured what he considered to be just the right place. It's time to meet your maker: an alkaline hydrothermal vent.

LIFE SPRINGS ETERNAL

'And what,' you may ask, 'is one of those?' The simplest answer would be a hot spring on the seabed, caused by a chemical reaction between sea water and a common mineral, olivine. Olivine – a greenish crystal made of iron, magnesium, silicon and oxygen – reacts with sea water to form serpentinite. And the striking thing about the booking hall here at Euston is that it is paved with serpentinite marble, a green stone with white serpent-like marbling, a fact that the fatalist in me can't help feel is significant.[7]

7 OK, cards on the table. I am not entirely sure that what builders call 'serpentinite marble' has anything to do with serpentinite the mineral other than a similarity in appearance, which is to say both are green with a serpent-like pattern. In fact, I'm not altogether sure 'serpentinite marble' has got anything to do with marble. It's probably made in a factory from brick dust. Frankly when you're looking for synchronicities you take what you're given.

Anyway, back to the plot. When sea water reacts with olivine to form serpentinite, it releases a great deal of heat, and this hot, mineral-rich fluid then rises as alkaline springs on the seabed. When it meets the cold sea water at the bottom of the ocean, the minerals precipitate out, like the limescale in a kettle, producing vents of porous white limestone. One famous example is the so-called Lost City in the middle of the North Atlantic, a ghostly hoard of some thirty gnarled chimney stacks, the tallest of which is some twenty storeys high.[8]

It's in the microscopic pores of this limestone, Mike Russell realised, that life may have got its start. The clue came from the way that all single-celled organisms store energy. Essentially, they act like tiny electric batteries, pumping protons across their membranes so that their insides become less positively charged than their outsides.[9] We call this a proton gradient. If energy is required for a chemical reaction somewhere within the cell, a proton is allowed to fire back through the membrane. The energy of this proton is then harnessed to create a molecule of energy-carrying ATP,[10] which then carries it to wherever it is needed.

Which begs the question, 'why bother?' Why go to all the trouble of storing energy in a proton gradient, instead of just making ATP straight away? Maybe, thought Mike Russell, it's a hangover from an earlier stage of life. McDonald's sells hamburgers in 119 countries, but banks its profits in dollars

8 In fact, the only example. Discovered in 2000, close to the Mid-Atlantic Ridge, the joint between the Eurasian and North American plates, at 30° north.

9 The acidity or alkalinity of a liquid is defined by its pH, being the logarithm to the base ten of the number of hydrogen ions (protons) per cubic metre.

10 Once it has given up its energy, ATP is converted to ADP, or adenosine diphosphate. It's not as complex as it sounds. ATP is basically a blocky carbon molecule with three phosphate groups getting on each other's nerves, so much so that one is happy to whizz off energetically leaving the other two behind.

because its first store was in San Bernardino, California. Maybe single-celled organisms bank in protons because that's the way it was done back in the 'hood. And one place you are sure to find a proton gradient is in an alkaline hydrothermal vent.

Why? Because 'acidity' and 'alkalinity' are just 'proton concentration' by another name. Acidic fluids have a lot of protons; alkaline fluids have few. The present-day oceans are slightly alkaline,[11] but thanks to a much higher concentration of atmospheric carbon dioxide, back in the day they would have been much more acidic.[12]

EXAMINE YOUR PORES

With that in mind, let's zoom in on one of the pores in the limestone chimney of a primordial alkaline hydrothermal vent and see what's going on. Amazingly, it's full of microscopic bubbles. Each one acts like a tiny battery. Inside, we've got warm alkaline fluid. Immediately outside, we've got acidic sea water. And betwixt the two we have a thin gel-like membrane made of iron sulphide, with a proton gradient across it.

11 Thanks to anthropogenic carbon dioxide, this is changing, and the world's oceans are becoming more acidic. In fact, their pH is changing ten to a hundred times faster than at any time in the past fifty million years. This will have positive and negative effects on marine life; we are all hoping that the positive outweighs the negative.

12 Ah yes – I forgot to mention. In the years since the Miller–Urey experiment, we've learned that, far from being full of ammonia, hydrogen and methane, the early atmosphere was full of oxidised gases. How do we know this? Because of zircon crystals. The oldest are dated at 4.4 billion years old, and their chemical composition indicates that they were formed at moderate temperatures in water, in an atmosphere full of nitrogen, sulphur dioxide and carbon dioxide. No oxygen of course; that came roughly 2.3 billion years ago, as a by-product of oxygenic photosynthesis. More about that in a moment.

Could such a bubble be the place that some prototype of life set up shop, using a natural proton gradient to drive chemical reactions? As we saw in the last chapter, to organise matter we need to be able to do work. Thanks to the proton gradient across its membrane, a bubble in an alkaline hydrothermal vent has energy on tap. What's more, its confined space is also a great place to concentrate chemicals. The concentration problem and the energy problem have been solved in one fell swoop.

And those aren't the only things these kinds of vents have going for them, because they are also rich in another vital constituent of living cells: transition metals. Cast your mind back to the periodic table, and you will remember that the middle of it is made up of a block of colourful, dense, mildly reactive metals with dependable names like iron, nickel, copper and zinc. Ever wondered why these are recommended as part of a balanced diet? Crucially, it's because we find them embedded within a bewildering number of proteins.[13]

Why might that be? Well, the thing about transition metals is that they make great catalysts.[14] Essentially they are wealthy philanthropists, with more electrons than they rightly know what to do with, and are happy to donate a few to the needy, safe in the knowledge that they will

13 Roughly a third of all proteins contain an embedded transition metal.

14 Chemists refer to an inorganic element's 'oxidation state', being the number of electrons a neutral atom has either gained or lost. Iron is particularly obliging, and is happy to gain one or two electrons, or lose as many as six. We then say it is capable of oxidation states −2 to +6. For organic molecules, we define oxidation state slightly differently. C-C bonds are neutral, C-H bonds reduce the oxidation state by −1, and non-carbon atoms increase it by +1. The carbon in carbon dioxide (CO_2) for example, is fully oxidised at +2; the carbon in methane (CH_4) is fully reduced at −4. In an oxidising atmosphere such as on Earth, methane is therefore chemically unstable compared to carbon dioxide. In a reducing atmosphere – in other words, one rich in hydrogen – it would be the other way around.

regain them somewhere further down the line. Alcohol dehydrogenase, for example, the enzyme in the liver that breaks down alcohol, contains a socking great zinc ion right in the middle of it, as do some three hundred other known enzymes. So far as we can make out, transition metals were the first catalysts, and were later enslaved by enzymes.

Membranes, proton gradients and transition metals: all of them make a good case for alkaline hydrothermal vents as the cradle of life. Add to that the fact that they are tucked away on the seabed, out of reach of the 700 million year asteroid bombardment that rattled the newborn Earth, and I hope I've got your attention. So how was the trick done? How do you get from something non-living to something living?

MAKE MINE A SINGLE-CELLED ORGANISM

Knowing that life is a symphony written in long-chain carbon molecules, it's clear that we need a source of carbon. What better place to get it than from the carbon dioxide dissolved in the primordial ocean? In long-chain molecules we most often find carbon bonded to hydrogen, so we are going to have to find a source of that as well. What about water? We've got plenty of that to hand.

No dice. It's possible to get carbon dioxide to react with water to produce long-chain carbon molecules and free oxygen, but it's far from easy. Plants do it, but they use sunlight, harnessed by a convoluted chemical pathway called oxygenic photosynthesis. That particular party trick didn't emerge until something like 2.8 billion years ago at the very earliest.[15] No, water won't do. So where are we going to get our hydrogen from?

This is where another feature of alkaline hydrothermal vents comes to the fore; serpentinisation produces lots of dissolved hydrogen. Carbon dioxide and hydrogen will react to make methane and other long-chain molecules, but you need both energy and a catalyst to get things going, much in the same way that you need a match and a fire lighter to get a decent fire to take in a grate. As we know, not only do our bubbles have energy on tap, but their membranes are made from iron sulphide, a catalyst which is perfect for the job.

One of my favourite Monty Python sketches is from *The Life of Brian*. 'All right, all right,' says a revolutionary John Cleese, 'but apart from better sanitation and medicine and education and irrigation and public health and roads and a freshwater system and paths and public order ... what have the Romans ever done for us?' After my chat with Nick Lane, I feel the same way. Apart from the protection from meteorites, the membranes, the transition metals, the proton gradients, the iron sulphide catalysts, the dissolved hydrogen and the carbon dioxide ... what have alkaline hydrothermal vents ever done for us?

PUSHING CARBON UPHILL

Of course it's a long way from a few smallish carbon molecules in an iron-sulphide bubble to a single-celled life form. Unlike the primordial soup theory, however, which

15 We currently think that oxygenic photosynthesis evolved around 2.7 bya in a single-celled organism called cyanobacteria. By 2.3 bya the build-up of oxygen in the atmosphere became significant, producing the Great Oxidation Event.

produces life like a rabbit out of a hat, our vent-based proto-life can make its way there in stages. One of Mike Russell's triumphs has been to show that the iron sulphide membrane of an individual bubble is permeable. That means that small molecules can escape, but larger ones are trapped, ready to undergo further reactions.

In the broadest of terms, the reaction of carbon dioxide with hydrogen would first produce small carbon molecules like methane, formate and acetate,[16] all of which would be allowed in and out of the membrane. In the next stage, these small molecules would react together to form medium-sized molecules such as amino acids and nucleobases. These would be trapped by the membrane, the proverbial fish in a barrel for further reactions which would then produce larger and larger molecules.

A crucial step would have been the creation of the first long-chain carbon molecule that was capable of making copies of itself, or, as we say in the jargon, self-replicating. In present-day organisms, this role is played by DNA, but it's unlikely to have been the molecule of choice for the very first life. For a start, DNA has a complex double-stranded structure, being a sort of 'twisted ladder', with 'sides' made of ribose and phosphate groups, and 'rungs' made of pairs of nucleobases.[17] In fact you can think of a DNA molecule as more or less being two RNA molecules

---

16 I know you're gagging for some chemical formulae here, so methane is CH_4, formate is HCO^{2-} and acetate is CH_3CO^{2-}.

17 This rigid structure makes it very stable. One of my favourite science facts is that real, working DNA has successfully been recovered from the toe bone of a 130,000-year-old Neanderthal found in a Siberian cave, enabling the sequencing of the Neanderthal genome, and making *Jurassic Park* seem like a real possibility. More about the Neanderthals later.

fused together,[18] a simple fact that has led many to suppose that RNA came first and later evolved into DNA.

Supporting this is the fact that DNA is essentially passive. To decode it, translate it into a recipe for amino acids, then assemble those amino acids into proteins requires RNA. Add this to the fact that RNA is able to write code into DNA, and catalyse reactions, and you start to build a picture of a busy chef who has decided to write his favourite recipes in a cookbook. Meaning that RNA is the chef, and DNA is the cookbook. All of which hints tantalisingly at an earlier epoch of life, predating DNA, when RNA ruled the kitchen. Evolutionary biologists call this the 'RNA World'; life, but not as we know it.

The grail of researchers like Lane is to create RNA from scratch in a model alkaline hydrothermal vent. At the climax of my visit, he leads me into a pristine lab, where a glass cylinder the size of a bricklayer's thermos flask sits on a bench top, trailing wires and surrounded by electronic monitors. It looks like a lava lamp on life support. I peer through the glass, not really sure what I'm expecting to see. Frankenstein's Molecule, perhaps? Could it be possible that, inside this sterile looking experiment, new life is taking its first shuffling steps?

A JUMBO IN A JUNKYARD

One of the most famous critiques of the primordial soup theory was made by Fred Hoyle. In his 1981 book *The*

---

18 There are two main differences. Where RNA employs the nucleobase uracil, DNA uses its close relative thymine, and the ribose groups of DNA have had an oxygen atom removed when compared with those in RNA. Hence the names RiboNucleic Acid and DeoxyriboNucleic Acid.

Intelligent Universe, he professed bemusement that something as complex as a single-celled organism could come about by chance. With characteristic Yorkshire phlegm, Hoyle put it this way:

> A junkyard contains all the bits and pieces of a Boeing 747, dismembered and in disarray. A whirlwind happens to blow through the yard. What is the chance that after its passage a fully assembled 747, ready to fly, will be found standing there? So small as to be negligible, even if a tornado were to blow through enough junkyards to fill the whole Universe.

Can a proton gradient across the membrane of an iron sulphide bubble trapped in an alkaline hydrothermal vent achieve what a tornado in a junkyard cannot? After my encounter with Nick Lane, I'm in the 'yes' camp. Most importantly, unlike the tornado, the vent doesn't have to make the entire cell in one go; it can do it in incremental steps. First, it just needs to do something simple, like take one molecule of carbon dioxide and react it with hydrogen to make methane. Next it synthesises the building blocks: amino acids and nucleobases. What's more, any large molecules that form can't ever leave, trapped as they are within gel-like bubbles of iron sulphide, encouraging them to form ever longer chains: handy stuff like proteins, and nucleic acids.

And here's the crunch. Far from being a shot in the dark, life is a slam dunk. We should expect to find it anywhere there's a hydrothermal vent bubbling alkaline vent fluid into an acidic ocean, and such vents are a feature of all newborn, volcanic, wet, rocky planets. Far from being a statistical fluke, life is just the chemical pathway by which carbon dioxide reacts with hydrogen to form methane. Or, as Mike Russell succinctly puts it: 'the meaning of life is to hydrogenate carbon dioxide.'

FIRST TO THE PARTY

Can that be true? Certainly life got started very quickly on Earth. To see just how quickly, let's remind ourselves of how planetary systems form. First, a shockwave from a supernova creates pockets of high density in surrounding gas and dust. These pockets then collapse under gravity to form clusters of new stars. As each new solar system condenses, it spins faster and faster, flattening into a disk, much in the same way that a skater spins faster as she draws in her arms.

Out in the disk, planets begin to clump together under gravity. The temperature of the protostar determines what type of planet forms where. Close to the star is the rock line, where the temperature is cool enough to allow rock to solidify. Here's where we find small rocky planets like Mercury, Venus, Earth and Mars. Further out is the snowline, beyond which water, methane and ammonia all freeze, and we find giant ice planets like Uranus and Neptune. In between are the giant gas planets, like Jupiter and Saturn.[19]

Eventually, the temperature and pressure of the protostar become so great that it 'switches on', and begins nuclear fusion. A blast of charged particles strafes the newborn solar

19 Rocky planets tend to be small both because there's not much rock to go round, and because their orbits are small so there's a shorter path over which to hoover it up. The gas planets like Jupiter are in the sweet spot, where there's lots of material and a large orbit. The ice planets have even larger orbits, of course, but not so much material, so tend to end up somewhere in the middle. Beyond the ice planets, orbits are enormous but material is even harder to come by, hence dwarf planets like Pluto.

system, blasting away the remaining gas and dust and leaving naked planets. For the first time, their home star lights the horizon. By this time the solar system is barely fifty million years old. Another fifty million years on, and the Earth is much like it is today, with a carbon-dioxide rich atmosphere, little land and an acidic ocean.

And here's the kicker. Fast-forward a mere 300 million years, and the organism that we call LUCA, the Last Universal Common Ancestor of all life on Earth, is eking out an existence in an alkaline hydrothermal vent. It uses the proton gradient between vent fluid and sea water to hydrogenate carbon dioxide – meaning to replace one or both of its oxygen atoms with hydrogen atoms – releasing energy.

As we know, LUCA wasn't the first life. It was the product of millions of years of evolution, one tier of which had probably been RNA-based. Other kinds of worlds almost certainly predated these RNA worlds, but their self-replicating molecules are lost to us. However you dice it, 300 million years was not a great deal of time to produce something as complex as LUCA. Life isn't rare, at least not in its proto-cellular form. It works straight out of the box.

Yet LUCA, as I've hinted, hadn't yet left the safety of the vent. To do that, it needed to develop a membrane capable of generating its own proton gradient. Was that a roadblock on the path to complex life? It would seem not. This may come as a surprise, but there's growing evidence that LUCA left the vent not once, but twice.

POPPING THE EVOLUTIONARY HOOD

Ever since Darwin sketched his first 'tree of life', expressing his idea that all life on Earth has evolved from a common ancestor, biologists have been arguing the toss over which species begat which. Formally known as taxonomy, the guiding principle of this somewhat fraught discipline was to try and group organisms according to their physical traits.

The name of the game was to divide creatures into groups that shared the same characteristics, and then to rank those groups in some sort of evolutionary order. On one level, of course, this makes complete sense. The grand sweep of evolution can be more or less summarised as a progression from the simple to the complex, so you'd think that you'd be able to sift and sort organisms into some kind of time line. At one end, you'd have the simple stuff, like single-celled bacteria, and at the other you'd have the complex, multicellular things such as woolly mammoths. Get into the fine detail, however, and it's a different matter.

For a start, the more closely related two species are, the greater the similarity in their outward appearance and the harder it is to rank them. A difficult job isn't made any easier by the fact that traits can just as easily evolve out as evolve in. While the grand sweep may be towards complexity, on a shorter timescale there is a great deal of ebb and flow. Humans are a classic example. Neanderthal man, who lived alongside us in northern Europe only 39,000 years ago, appears to have had a bigger brain than we have, and may have been more intelligent than we are. If an alien taxonomist arrives on a barren Earth some million years hence, and finds a human and a Neanderthal skull, he could be forgiven for assuming that the Neanderthal version was the more recently evolved of the two.

And if that weren't bad enough, there's convergent evolution to deal with. As we shall see in much more detail in

Chapter Seven, there are some traits which crop up time and time again, as nature evolves similar solutions to similar problems. Eyes, for example, have evolved independently scores of times. Both humans and octopuses, for example, have camera eyes. If anything, octopus eyes are slightly better designed as they lack a blind spot. Unfortunately for the taxonomist, this means that organisms that look alike aren't necessarily close relatives.

All of these complications combined to make taxonomy one of the most frustrating, controversial and internecine disciplines ever created in science.[20] Once it became possible to examine the structure of DNA, however, all that changed. As we've already seen, DNA is a ladder-shaped molecule twisted into a helix, where the sides of the ladder are formed by alternating sugar and phosphate groups, and the rungs are made of the four bases, guanine, adenine, thymine, cytosine. The exact sequence of those bases stores all the information needed to create the organism from scratch, be it an amoeba or an ostrich.

That sounds complicated, but the way that DNA functions can be grasped simply by renaming the four bases G, A, T, C. To cut a long story short, the bases form a four-letter alphabet that can be used to make up two kinds of DNA. The first type, called 'coding regions', holds the recipes for all the proteins the host organism is made up of. Somewhere in your DNA there will be a coding region − or gene − with the recipe for haemoglobin, for example, which is the

---

20 It's worth knowing that before DNA sequencing, life was divided into five kingdoms: animals, plants, fungi, protists and prokaryotes. Within these various kingdoms, there were divisions into phylum, class, order, family and, finally, genus and species. As *Homo sapiens*, for example, we sit within the kingdom of animals, in the phylum of vertebrates, in the class of mammals, in the order of primates, in the family of the great apes, in the genus of humans, in the species of modern humans.

oxygen-carrying protein in the red corpuscles of your blood.

The second type of DNA, called 'non-coding regions', are a little more mysterious. They far outnumber the coding regions – about 98 per cent of your DNA is non-coding – and control how the coding regions are switched on and off, as well as acting as a kind of a junkyard where bits of code can be stored that might come in handy at a later date.

The entirety of an organism's DNA is known as its genome, and, since the invention of DNA sequencing by Fred Sanger in 1977, it has been possible to transcribe the bases for an entire organism. It all boils down to this. Whereas heredity used to be a matter of opinion, now it is a matter of fact. Effectively we can flip the hood of an organism and read off its genetic code, then compare it with the code of another organism. We can therefore see directly how the two are related, and construct a tree of life not from an organism's outward appearance, but from its genome. This new discipline is called phylogenetics, and it has transformed biology.

CHOOSE YOUR DOMAIN NAME

One of the first big surprises of phylogenetics was the discovery by the American biologist Carl Woese that a previously undiscovered branch of life had been hiding in plain sight. Sadly, I'm not talking about Bigfoot, which I'm fairly sure even a taxonomist would reveal to be a man from the Washington State tourist board dressed in a furry suit.[21] The branch of life that Woese discovered was altogether more modest in size. In fact, it was microscopic.

21 Is it just me who finds it a coincidence that the US state with the highest number of Bigfoot sightings is the same one where flying saucers were first spotted?

In a nutshell, Woese discovered that the kingdom everyone had been calling bacteria was actually two completely different kinds of single-celled organism. Although both creatures looked similar under the microscope, when it came to their genetic make-up they were about as different as it is possible to be. What's more, neither was clearly the ancestor of the other; both appeared equally ancient.

At the time it was thought that there were only two kinds of cell on Earth: those with a nucleus, named eukaryotes – Greek for 'true kernel' – and those without, named prokaryotes, as in 'before kernel'.[22] As the names suggest, it was believed that the prokarya had preceded the eukarya. Woese's discovery was that prokarya were actually of two types, as distinct from one another as they were from eukarya. He proposed a whole new classification for living organisms, dividing them into three domains: the eukarya, the bacteria and the archaea.

One of the striking differences between the bacteria and the archaea is in the structure of their cell membranes. In both cases, the building block is a lipid with a water-loving phosphate molecule at one end, but, in the case of bacteria, the lipid is a fatty acid, whereas in the archaea it's an isoprene.[23] If life began in a primordial soup, how could this have come about? If there are two kinds of single-celled organisms with different cell membranes, aren't we asking for a junkyard tornado to assemble a jumbo jet not once, but twice?

Alkaline hydrothermal vents, of course, provide us with a possible answer. LUCA lived in a vent, and it needed to

22 Going back to the five kingdoms: animals, plants, fungi and protists are all eukaryotes. In case you're wondering, protists are single-celled eukaryotes, the most famous example being the amoeba.

23 Fatty acids are basically zigzag chains of hydrogenated carbon. Isoprenes are, too, but they also have side branches, also made of hydrogenated carbon atoms.

evolve a membrane in order to leave. And it did. Twice. One iteration gave rise to the bacteria and the other to the archaea. Independently, each evolved its own proton pump, a nanomachine in its cell membrane capable of recreating the proton gradient that had powered the metabolism of LUCA. Both domains still use this mechanism today, storing energy by pumping protons across their membranes, then allowing them back through in order to generate ATP.

So far, so good. Given an alkaline hydrothermal vent, LUCA is easy to make. So are membranes; so easy they evolved twice. Is this the case all the way down the line? Is every step along the path from single-celled life to technologically advanced civilisations the evolutionary equivalent of a cascading line of dominoes? If you are hoping the answer is yes, I am here to disappoint you. The creation of intelligent life appears to have hinged on one extraordinary event. If you're new to biology, all I can say is you are in for a shock.

A PLAGUE ON BOTH YOUR HOUSES

To summarise where we've got to so far, we've learned that, far from being the statistical equivalent of a Boeing assembled by a junkyard tornado, the last common ancestor of all life on Earth arose swiftly by a series of high-probability steps. Our best guess is that it set up shop in iron sulphide bubbles, sheltered in the limestone chimney of an alkaline hydrothermal vent. It set sail into the ocean with a brand new membrane on at least two occasions, and our present-day bacteria and archaea are the direct descendants of these two rival species. Crucially, what happened next was . . . nothing.

The sad fact for fans of intelligent life is that the bacteria and the archaea have remained single-celled and, well, dumb, from their inception right up until the present day. For four billion years they have resolutely avoided evolving into anything remotely resembling a complex multicellular organism, let alone a technologically advanced intelligent one.

In truth, they haven't really needed to. Whatever we humans might want to believe, the world belongs to archaea and bacteria. They are, without question, the most successful organisms on the planet, bar none. Even in our own bodies, they outnumber our cells ten to one. We find them in the ocean, the atmosphere, salt lakes, even in the reactors of nuclear power plants. Almost anywhere there is a source of energy, bacteria and archaea have found a way to exploit it, but never in order to become more complex. All their evolution has been biochemical; structurally, they remain simple bags of chemicals, reproducing, feeding, excreting waste, dying, and doing little else.

Not that they haven't left their mark on the planet. One of their most striking contributions has been to excrete oxygen into the atmosphere, a gas completely absent from the primordial Earth. We find the first evidence of free oxygen at around 2.3 billion years ago, over a billion and a half years after LUCA's bubble burst and it left the vent. Nicknamed the Great Oxidation Event, this marks the bacteria's discovery of a sophisticated biochemical pathway called oxygenic photosynthesis, the harnessing of sunlight to rip electrons from water and force them on to carbon dioxide, fixing carbon and releasing oxygen as a waste product.[24]

Oxygen, as we all know, is reactive, and once it entered the atmosphere it rapidly set about bonding with anything

---

24 Other kinds of photosynthesis came first. In fact, one of the first indications we have of life on Earth are so-called 'banded iron' formations, dated at 3.2 billion years old, caused by an early kind of photosynthesis that used iron instead of water.

it could get its needy little orbitals on: all too soon it was rusting metals, oxidising salts and converting atmospheric methane into carbon dioxide. Methane is a potent greenhouse gas, and removing it had a profound effect on climate. In fact, it's believed that falling methane levels were a significant factor in triggering the Huronian glaciation, when the global temperature dropped dramatically, causing runaway growth of the polar ice caps to the extent that the entire planet froze over in what's known as a Snowball Earth.

And that might have been that, but for one extraordinary event that was to change the entire trajectory of life on Earth. Without it, we'd still be living in a world of micro-organisms. There would be no animals, no plants, no fungi and no amoebae. Few living things would be visible to the naked eye other than the odd bacterial colony. So far as we can tell, this extraordinary event happened only once, and it created a new kind of cell, the eukaryote. Unlike the bacteria and the archaea, eukaryotic cells are highly organised and have a nucleus. Much of the origin of these extraordinary cells remains a mystery, but this much we know; they were created by the enslavement of a bacterium by a hungry archaeon.

MIGHTY MITOCHONDRIA

There are two kinds of prokaryotes in this world, those that create their own food from inorganics, called autotrophs, and those that feed by eating other cells, called heterotrophs. Either way, the goal is to end up with glucose, which can then be burned to create energy in a process called respiration. Usually that means shoving it into something called the Krebs cycle, a repeating chain of biochemical reactions which pump protons across the cell membrane before allowing them back through to generate ATP.

Why am I telling you this? Because something like one and a half to two billion years ago, a heterotrophic archaeon – that's singular for archaea, by the way – ate an autotrophic bacterium. Or at least it tried to. It engulfed it and attempted to digest it, no doubt intending to break it down into sugars that it could then shove into its very own Krebs cycle. Thankfully for us it failed.

Instead, the bacterium survived. In fact, it more than survived; it thrived. Together, the two organisms negotiated a delicate pact. The bacterium became a permanent fixture within the archaeon, forming what biologists call an endosymbiosis. In return for shelter and a ready supply of glucose, the bacterium used its Krebs cycle to supply the archaeon with ATP. In effect, it became a power plant for its host. Why was all of this so groundbreaking? The reason, as ever, is to do with energy and information.

Because bacteria and archaea store energy by pumping protons across their membrane, they have a problem. The bigger they get, the less membrane they have relative to their mass, and the less efficient they get. You can't have more than one membrane, but you can have as many slave bacteria as you like. By outsourcing respiration to an army of supplicants, the archaeon was able to generate an extraordinary amount of energy. A whole new level of complexity became possible. The bacterium became a mitochondrion and the eukaryotic cell was born.

As I hope you can see from my drawing over the page, they were a world apart from their prokaryotic counterparts. Where bacteria and archaea are one-horse towns, the eukarya are sparkling citadels, full of eye-catching new structures. We've already mentioned the power plant that is the mitochondrion, a

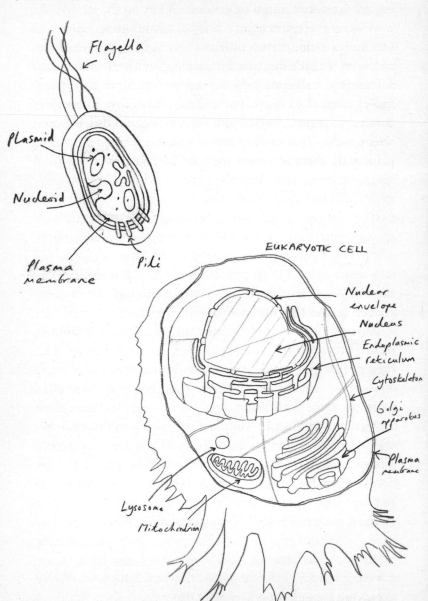

PROKARYOTIC CELL

Flagella

Plasmid

Nucleoid

Plasma membrane

Pili

EUKARYOTIC CELL

Nuclear envelope

Nucleus

Endoplasmic reticulum

Cytoskeleton

Golgi apparatus

Plasma membrane

Lysosome

Mitochondrion

stripped-down autotrophic bacterium whose job it is to supply energy to the cell; and the copyright library, in the form of the nucleus, that could now house an almost limitless quantity of DNA. In addition, these eukaryotes have a factory where RNA can assemble proteins, called the endoplasmic reticulum, and a UPS service called the Golgi apparatus which can package up those proteins for export. There's a road network in the shape of the cytoskeleton, a web of pathways throughout the cell along which metabolites can be transported, and a waste-processing plant in the form of a lysosome. Life 2.0 had arrived.

GIVE ME SOME OXYGEN

The stage was now set for complex life. When the last of the Snowball Earths turned to slush around 635 million years ago, the first multicellular creatures made a tentative exploration of the newly warm shallow oceans.[25] We call this period the Ediacaran, and it saw the rise of some truly bizarre new life forms as well as a dramatic increase in the level of oxygen. If you want to catch a glimpse of beings that really look alien, you need to Google the Ediacaran biota.

That name 'biota' – meaning 'living part of a biosystem' – is well chosen, because in many cases you'd be hard pushed to say whether the Ediacaran fossils we have are made of sponges, plants, fungi, jellyfish or something else entirely. The earliest appear to have been large disc-like creatures, rooted to the ocean floor at considerable depths.

---

25 We used to think the increase in oxygen 'released the brakes' on evolution, and enabled the evolution of complex life. We now believe it was the other way around; the evolution of the Ediacaran biota helped oxygenate the oceans and sea bed through filter-feeding and burrowing.

They don't have mouths or limbs, and we can't tell if they had internal organs. Our best guess is that they lived alongside microbial mats, absorbing nutrients through their skin.

Unfamiliar as they are, however, these strange creatures represent a crucial step towards intelligent life. Not only does size bring security – it's hard for a heterotrophic micro-organism to swallow a sponge – but it also brings greater energy efficiency. Add to that the fact that burning glucose with oxygen produces much more ATP than fermenting it, and you can start to see how much more energy was becoming available to drive complexity.[26]

The end of the Ediacaran saw a boom in complex life not seen before or since. Almost overnight, a fresh clade of exotic creatures swept the globe. We call them the metazoans, or animals. Fittingly, we call this extraordinary radiation the Cambrian Explosion, and its resonance can be seen in all intelligent life today. Again, their key innovation was to do with energy. I'll call a spade a spade: life evolved a mouth, a gut and an anus.

LET'S HAVE US A BILATERAL TRIPLOBLAST!

High in the Canadian Rockies, in British Columbia, there's a small limestone quarry shot through with a seam known as the Burgess Shale. This is the site of arguably the

---

26 Because there were relatively low oxygen levels (roughly 1 per cent atmospheric) when the eukaryotic cell came into being, it seems likely that the very first mitochondria fermented glucose rather than burning it in oxygen. The progenitor of endosymbiont theory, Lynne Margolis (ex-wife of Carl Sagan, by the way), had a different view. She believed that the first mitochondria were aerobic (oxygen-burning) rather than anaerobic (fermenting), which is why the host archaeon liked them so much, given that the oceans were filling with free oxygen. Who's right? As my father used to say, you pays your money and takes your choice.

most important fossil find in history, the palaeontological equivalent of Pompeii. Five hundred and forty-one million years ago, right at the beginning of the Cambrian, a mudslide into shallow water engulfed a profusion of bewilderingly diverse life forms. The fossils they left behind are exquisitely preserved, as if a trawler's net had been lowered into the prehistoric ocean and hauled on to the deck.

That said, not one member of this once-in-a-lifetime catch looks like anything you would want to deep-fry and eat with chips. Opabinia, for example, is like a slug in a ball gown with a single ominous pincer. Aysheaia looks like a roll of lino with a mouth at one end, and Marrella resembles nothing so much as an overcreative trout lure. Yet, strange as they first appear, some of these long-extinct creatures share a basic body plan which we humans have inherited. In short, they are bilateral triploblasts[27] and they are the proud owners of a digestive tract.

That last bit is important, because – as you might have guessed – a digestive tract is yet another way to up the energy stakes. What better way to make a living than by shoving whole organisms in at one end, shredding them with teeth, digesting them with enzymes in the gut, then ferrying the resultant glucose to your mitochondria to generate swathes of valuable energy? Not to mention the rather satisfying moment when you egest everything you don't want in a lazy dropping.

No one is exactly sure what became of the sightless and mouthless Ediacarans, but it's a fair guess that many of them spent their final hours in the gut of a bilateral triploblast. With the ability to hunt and eat, the stage was now set for this particular brand of multicellular life to reap even more energy from the oceans, increasing its complexity along the

27 Bilateral, as you might guess, means their left side mirrors their right side. The triploblast bit refers to the fact that they have a body cavity arranged about their digestive tract.

way. Most of you will know the romance which follows. Like all the best stories, it divides neatly into three acts. In the first, life is established on land. The second sees the rise of the dinosaurs. And the third, in which we are lucky enough to be living, sees the ascendance of mammals.

THE AGE OF FISH AND PLANTS

The Palaeozoic Era, which runs from the Cambrian Explosion to the Great Dying at the end of the Permian, spans roughly half the lifetime of complex life. Although there were no more Snowball Earths, the planet still endured two Ice Ages.[28] The main events were the evolution of land plants and the subsequent evolution of fish into amphibians, and then into four-legged land animals, known in the trade as tetrapods. By the end of the Permian, the sixth and final period of the Palaeozoic – see my handy guide to geological periods on pages 190–191 – there were trilobites and fish galore on the ocean shelves, and horrendous snaggle-toothed ancestral mammals called Gorgonopsids roaming the forests, the largest of which was the size of a grizzly bear.

Not that you are ever likely to meet one, because, as the name suggests, the Great Dying is the largest of the known Big Five extinctions.[29] When you hear Al Gore talk about global warming, this is exactly the kind of nightmare scenario he is worrying about. At the time the continents were all joining together in one huge geological love-in called

28 The Andean-Saharan lasted thirty million years, and spanned the Ordovician–Silurian extinction. The Karoo Ice Age lasted roughly sixty million years, and took place during the Carboniferous.

29 I know you want to know, so these are: the Ordovician–Silurian (the former pronounced "Ordo-vee-shan"), the Late Devonian, the Permian–Triassic, the Triassic–Jurassic, and the Cretaceous–Paleogene.

Pangea, and fissure-like volcanoes produced a carpet of lava the size of continental Europe. The colossal amounts of carbon dioxide released caused a runaway greenhouse effect, and the mean global temperature rose by some six degrees.

Ocean currents, as you may know, rely on ice at the poles to sink cold oxygenated water, which then wells up in the tropics. When the polar ice melts, as it did during the Great Dying, these currents switch off and tropical waters become less oxygenated. That means curtains for oxygen-loving marine life. Not only that, but the centre of Pangea became a barren desert, eradicating swathes of newly evolved land species. All told, at the end of the Palaeozoic a staggering nineteen out of twenty species went extinct. Think about that the next time you fill up with diesel.

THE AGE OF THE DINOSAURS

The following era, the Mesozoic, also ended in a mass extinction.[30] Thanks to the unbounded interest of children the world over, these warm geological periods are the ones we know best: the Triassic, Jurassic and Cretaceous. The Triassic essentially saw the recovery of life after the Great Dying, only to be followed by what many believe was a meteorite impact at the Triassic–Jurassic boundary. The resulting extinction enabled the rise of the dinosaurs, and also saw the emergence of our direct ancestors, the

---

30 By now you may be getting the picture; geological periods are often defined by extinctions. Typically reddish oxygenated rocks like limestones and sandstones are followed by black, unoxygenated ones like slates and shales. They tend also to be named after the places where they were first discovered: the Cambrian, for example, is Latin for Wales; the Silurian is named after the Welsh tribe the Silures. The Jurassic gets its name from the Jura Mountains, close to CERN in Switzerland. And the Devonian is named after, well, Devon.

Recent Geological Time

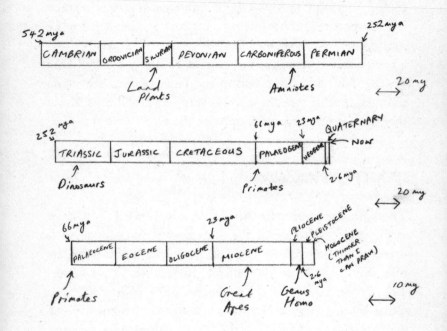

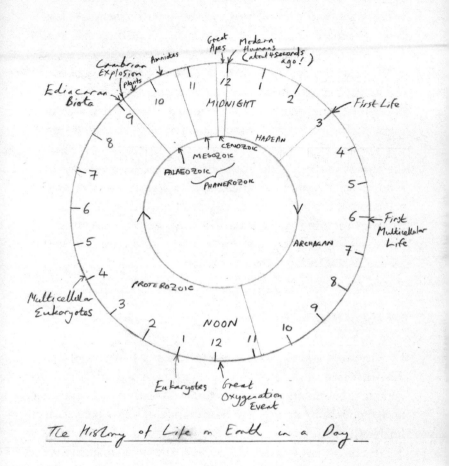

The History of Life on Earth in a Day

mammals. By the time of the Cretaceous, mammals had diverged into two main groups: those which gestated their young in abdominal pouches and those with placentae. It is from placental mammals that primates, and therefore we, are descended.

We now find ourselves at the boundary of our present era, the Cenozoic. As already mentioned, the Mesozoic ended with a bang some sixty-six million years ago, when a giant space rock some ten kilometres in diameter slammed into the Yucatán Peninsula in Mexico, releasing a billion times the energy of the atomic bomb that fell on Nagasaki. When random catastrophes like that occur, it tends to be the large predators at the top of the food chain that suffer. In this case that was the dinosaurs, of which only the birds survived.[31] As we all know, our ancestors the placental mammals were all too happy to step into the breach.

THE AGE OF MAMMALS

Evolution, as you can see, often proceeds by fits and starts. The first step of the cycle is a radiation, where all sorts of genetic experimentation goes on as organisms adapt to new niches. The next is the dominance of a few particular forms: the land plants, say, in the Silurian and Devonian.[32] Next comes an extinction, where the vast majority of species are wiped out on an utterly random basis. For the Gorgonopsids, that reckoning was the Permian extinction. The cycle is then free to begin again. And, true to form, a

31 Other large creatures also managed to sneak through what is called K-Pg extinction event, such as the crocodiles.

32 It's possible that the rapid expansion of land plants sucked up so much carbon dioxide that they caused the Karoo Ice Age in the Carboniferous.

radiation of reptiles in the Triassic saw the rise of the dino-saurs in the Jurassic and Cretaceous.

After the greenhouse world of the dinosaurs, the Cenozoic has generally seen a slow decline in global temperature, leading to the present Ice Age, the Quaternary, which began 2.6 million years ago. I know what you're thinking; if this is an Ice Age, you hadn't noticed, but an important feature of the Quaternary has been a gradual back and forth of polar ice. Epochs where the ice advances, known as glaciations, are interspersed with ones called interglacials where the ice retreats. As you might have guessed, we are in an interglacial right now, called the Holocene, where temperatures have been remarkably stable at roughly present-day levels for some 11,000 years.

But back to my point. It's thought that, as the polar ice formed, the planet became drier, and the African forests began to dwindle, giving way to savannahs.[33] Forests, as we all know, are the natural habitat of apes. Could that have been the selection pressure that encouraged our ancestors to leave the trees? Whatever the reason, some seven mil-lion years ago they began to diverge from the ancestors of chimps. Interestingly, the adaptation that helped them wasn't one that made them smarter. That came much later. They diverged because they could walk on two feet.

What good did that do them? Well, not much, so far as we can make out, for several million years. The fossil evidence tells us that for the following four million years they gradually got better at walking upright, but that was

---

33 After an initial hot flush, the Cenozoic has seen a mighty cooling. We think the Antarctic first developed glaciers around thirty-four million years ago, then began to cool dramatically once it became isolated from South America, roughly twenty-three million years ago, eventually becoming covered in ice fourteen million years ago. The Arctic took a little longer, and iced over about 3.2 million years ago.

about it. We find Ardepithecus shuffling across the forest floor some five million years ago, and Australopithecus sauntering upright on the savannah about a million years later, but they weren't the smartest customers, with individual brain capacities roundabout the same as that of their cousins the chimps.

Clearly they were doing something right, because just under two million years ago we find a radiation of hominins, or 'human-like' species, all living cheek-by-jowl in Africa.[34] Top billing goes to *Homo erectus*, with a brain capacity twice that of a chimpanzee, increased body size and smaller teeth. Not only was this species the first true hunter-gatherer, it was also the first tourist, with some members leaving Africa to take up residence in Asia. The 'smaller teeth' bit is particularly interesting, because it may hint at another link between energy and complexity. In short, there are some indications that *Homo erectus* could cook.

Although we have yet to find solid evidence that it could control fire, the fact that *Homo erectus* has small teeth, small guts and slept on the ground rather than in the trees are all intriguing clues. After all, cooking breaks down the long-chain carbon molecules in food, making it easier to chew and digest, dispensing with the need for a robust gut and dentition. It also means you can extract more energy from that vital adaptation, the gut. You can then expend that energy making your brain more complex, which in turn makes you a better hunter, creating a virtuous cycle.

There's another thing about cooking, of course. It's also a group activity, set around a campfire, which must have

34 Also around at the time were *Homo habilis* and *Homo ergaster*, as well as *Paranthropus robustus* and *Paranthropus bosei*.

encouraged all sorts of group bonding. As we shall see in the next chapter, intelligence is by no means confined to humans, but all the animals in which it occurs are social. It also encourages parenthood. Baby chimps are pretty much on their own once weaned, but cooking is technical, and not the kind of thing you'd ask a baby to do unless you were really in a rush. Cooking therefore encourages the dependency of children on their parents, buying time for childhood, a time of imaginative and creative development.

PULLING MUSSELS FROM A SHELL

To be fair, there's currently a great deal of disagreement about what *Homo erectus* could or couldn't do, and conclusive proof of tall stories by the campfire are a long way off. The conventional view is that they changed little in nearly two million years, using the same basic stone tools right up until their extinction some 140,000 years ago. In a book about communicating with aliens, however, I can't help but mention a recent and very controversial discovery.

This is also worth a Google. In scientific terms it's a scratch on a mussel shell, but it would take a bard to do it justice. Recently dated at 500,000 years old, it comes from a treasure trove of *Homo erectus* fossils found on the Indonesian island of Java by a Dutch surgeon named Eugene Dubois in 1891. Something – or someone – has carved a zigzag pattern on it, and, when you see it, it's hard not to feel a deeply human connection. I defy anyone to look at those markings, made by an ape nearly half a million years ago, and not to imagine that either it knew what beauty is, or sought order in a confusing world, or at the very least was bored and looking for something to do.

The next time in the fossil record we find anything like these patterns, they are scratched in ochre by anatomically modern *Homo sapiens*.[35] The site they come from, the Blombos Cave on the southern Cape coast, has been one of the richest sources of early human artefacts ever discovered. Dated at 100,000 years old, these criss-cross etchings speak to the existence of beings just like ourselves, capable of imagination, creativity and abstract thought. Something crucial is happening here. Information is no longer simply being stored in DNA. It is being stored in a network of brains. The name of that network, of course, is culture.

GONE WALKABOUT

After a genesis in Africa some 200,000 years ago, *Homo sapiens* rapidly spread across the Levant, heading for Asia and Europe. With a seemingly solitary leap, its brain size had increased by over a third in comparison to *Homo erectus*, and was now over three times the size of that of a chimpanzee. If art is a proxy for language, as many palaeoanthropologists believe, the indication from the Blombos Cave is that even these very first *Homo sapiens* were singing 'I don't know but I've been told' as they walked.

Or maybe they just chatted about the weather. Either way, 45,000 years ago they reached Europe, where they

---

35 We aren't completely sure how we are related to *Homo erectus*, but one of the most widely accepted hypotheses has it that the African branch of *Homo erectus* evolved into *Homo heidelbergensis*, which then spread to Europe and Asia where it evolved into *Homo neanderthalis*. African *Homo heidelbergensis* then evolved into *Homo sapiens*.

encountered the Neanderthals, a separate hominin line that had already been living there for some 150,000 years. A talented bunch with their own toolmaking and burial traditions, as already mentioned the Neanderthals had bigger brains than us and were arguably better adapted to the cold climate. Whatever happened, it doesn't seem to have involved much cooperation, because within 5,000 years the Neanderthals were extinct. Never let it be said that humanity doesn't have a dark side.

It's around this time that we detect what is often called the 'Great Leap Forward': a step change in human culture. There's evidence of elaborate burial rituals, of wearing animal skins as clothing, and using pit-traps to hunt prey. By the time we reach 40,000 years ago, you can hardly move in Europe or Asia without stumbling across all manner of cave paintings, jewellery, fishing hooks and flutes.

The invention of farming, roughly 11,500 years ago at the end of the last glacial maximum, secured a growing network of villages with a reliable energy supply. A few millennia later, around 3600 BC (5,600 years ago), came the founding of Sumer, the world's first civilisation, and, by 3100 BC, Sumerian had become its first written language. Now information could be stored not just in the brain, or in a network of brains, but as hard copy. And progress continues to be relentless. Today, a dwindling supply of energy from fossil fuels supports a worldwide population of seven billion souls, connected by a digital network that contains just about every single bit of information amassed by humanity to date.

THE UNBEARABLE LIGHTNESS OF BEING

So that's why we've never heard from aliens, right? The whole of human evolution hinges on one key event, the creation of the eukaryotic cell, without which the Earth would still, even four billion years later, be nothing but a petri dish for bacteria and archaea. There's nothing out there for us to talk to, because, although single-celled life is common, complex life is rare, and intelligent life rarer still. The series of flukes that led to intelligence in humans hardly makes communicable civilisations look like a dead cert.

So what are the chances that I should be here? Let's assume for a moment that the time in Earth's history when a given organism evolves is a good guide to its probability. This is a bit like rolling a dice once a minute. After six minutes, all things being equal, you'd expect to have rolled at least one six. Since the Earth has been rolling the dice for roughly four million years, and assuming the rate of mutation is roughly constant, we can calculate the probability of the various stages of life. Single-celled life has been there from the first roll of the dice, so its probability is 1. Eukaryotic cells emerged halfway through Earth's history, so their probability is ½, or 1 in 2. And we humans evolved 200,000 years before the present, so our probability is 200,000/4,000,000,000, or 1 in 20,000.

Of course that calculation is riddled with so many assumptions it's little more than a curiosity. The biggest is that the Earth is somehow typical of all life-bearing planets. To really get a feel for the probability of complex life, we need detailed knowledge of a large sample of life-bearing planets, something we are unlikely to have for at least a decade or two. Another is that only complex organisms with our own particular genetic make-up – that's to say bipedal hominins – possess the kind of intelligence necessary to evolve technology and communicate. But it doesn't look encouraging, does it? It suggests that simple life is common, but intelligent life is rare. From time to time a light comes on, but everywhere else is in darkness. Our project is in vain.

Well, not quite. Because, as we are about to see in the next chapter, it turns out that we humans are not quite as unique as we might think. There are other intelligent species on the planet, and an argument can be made that intelligence is just as common an adaptation as land-dwelling or flight. As we are about to see, there's a good chance that if we rewound the tape and played evolution out again, we'd find ourselves on a planet of the apes, dolphins or crows.

Not only that, but by focusing so sharply on the fine detail of evolution here on Earth, we've neglected the bigger picture. Yes, we are here because of a series of biological and climatic flukes. But we are also here because we had the time to evolve. We happen to live on a stable planet in a stable solar system, orbiting a quiet, long-lived star. When we look out into the galaxy, hoping to see our neighbours waving back at us, we are assuming that planets and solar systems like ours have existed since the Milky Way was founded, some thirteen billion years ago. But what if that's just not the case? What if life started on Earth at the same time that life could start anywhere in the galaxy? What if we are one of a whole host of civilisations that are just waking up? In order to climb evolution's ladder to complexity, we need to avoid its biggest snake: gamma-ray bursts.

HERE BE DRAGONS

In 1963, following the Nuclear Test Ban Treaty, the US launched the Vela satellite, essentially to make sure that what was then the USSR was playing ball. These high-flying satellites were tuned to detect the distinctive pulses of visible light, radio waves, x-rays and gamma-rays emitted by nuclear weapons tests. Instead, they found something else: bursts of pure gamma-rays coming from outside the solar system.

To begin with it was assumed that these bursts must be coming from somewhere within the Milky Way, but by the mid-1990s it was realised that they are in fact coming from distant galaxies, many of which are almost halfway across the observable universe. To be as bright as they appear in our sky, having travelled billions of light years to reach us, means that they must be extraordinarily intense. One particularly sobering fact is that a typical gamma-ray burst (GRB) contains as much energy as the Sun radiates in its entire lifetime, condensed into a pulse lasting just a few seconds.

We still aren't exactly sure what causes these blistering infernos. The shorter bursts are thought to be caused by the collision of binary neutron stars. The longer, more powerful bursts are thought to result from the gravitational collapse of colossal stars, known as hypernovae. One thing is certain, however. You don't want to be anywhere within 10,000 light years of one, or you'll be fried. The near side of any Earth-like planet would be toast, and the far side would then get blasted by a shower of secondary radiation. The ozone layer would be banjaxed, and any remaining life would then be decimated by the Sun's ultra-violet radiation.

We now have a lot of data on GRBs, and in 2014 two

astronomers named Tsvi Piran, of the Hebrew University of Jerusalem, and Raul Jiménez, of the University of Barcelona, crunched the numbers to find out what the risks are. Their findings make extremely interesting reading. For a start, they calculate that there's a 90 per cent chance that the Earth has been fried by at least one GRB at some time during the last 4.6 billion years, and a 50 per cent chance that it has been hit within the last half-billion. Could one of the Big Five mass extinctions have been due to a GRB?[36]

But it's their conclusions about the suitability of the universe to past life that really give you pause for thought. For a start, only 10 per cent of all galaxies have few enough GRBs to support life, and, even then, you'd better be nowhere near the centre where they tend to blow up more frequently. The Earth's position, some 25,000 light years from the centre of the Milky Way – which is, as you might have guessed, a member of that lucky 10 per cent – is now looking like prime real estate. But, most importantly, they calculate that life would have been impossible on any planet, in any kind of galaxy, anywhere in the universe before five billion years ago.

What does that mean? The bottom line: we can trace two imaginary spheres around the Earth. The first is five billion light years in radius, and within it we can expect to find single-celled life. Beyond that sphere the universe is barren, because any galaxy older than that is still being fried by GRBs. The second sphere is a billion light years in radius. If life on Earth is typical – and that's a fairly big 'if' – then within it we can expect to find technologically advanced societies. After all, it took four billion years for humankind

36 A prime candidate is the Ordovician–Silurian, for which there seems to be no convincing climate event or meteor strike.

to develop radio technology, and if that represents some kind of average, then the very oldest communicable societies will be a billion years ahead of us at most.

And finally, it raises a question. What if the Earth isn't typical, and the rise of our own technological society has been inordinately rapid? What if, on average, it takes six billion years to invent radio technology, rather than our own four billion? In that case, we have a chilling solution to the Fermi paradox: we are alone. And if it takes, say, an average of five billion years to evolve a communicable society? Well, then, it's all to play for. Maybe the galaxy isn't dead. It's asleep. And round about now it's going to start waking up.

CHAPTER SEVEN

Aliens

In which the author conjures real aliens, and learns that life is a servant of two masters. One, consciousness, hopes that the universe lasts forever. The second, cosmos, seeks an untimely end. No prizes for guessing which one is winning.

At first he assumed it had to be some kind of hoax. As Assistant Keeper of the Department of Natural History of the Modern Curiosities of the British Museum, George Shaw was often confronted with biological odd-ities, and a depressing number proved to be illegitimate. Accompanying the pelt was a sketch by its donor, Captain John Hunter, governor of the newly founded colony in New South Wales, purporting to show the animal when alive. Could the drawing be a fake, too? Even by Antipodean standards, the creature that confronted him was bafflingly bizarre.

Where to start? Its tail was flat, like a beaver's; its body was like that of an otter, or maybe a mole. Or, come to think of it, a seal; yes, he thought, that fur resembled nothing so much as a seal pelt. Like a seal, each of its four legs was home to a webbed foot. Yet search as he might, he could find nothing on the creature's abdomen that in any way resembled a nipple. As proclaimed by the Swedish zoologist Carl Linnaeus in his definitive *Systema Naturae*, the class of *Mammalia* was reserved for 'animals that suckle their young by means of lactiferous teats'. But if it wasn't a mammal, what was it? Could it possibly belong within the class of *Amphibia*, many of whose members 'appear to live promiscuously on land or in water'? Yet who had ever heard of an amphibian with fur?

And then there was the business of the creature's rear ankles. On each was a pronounced spur, much as you might see on a champion cockerel at a country fair. And that was the least striking of the creature's avian characteristics.

To be blunt, it had the head of a duck. Or, at least, the mandible of a duck; there was something distinctly fishy about its minute eyes, so small that one had to hunt to find them amidst the tufts of fur. Shaw was well acquainted with the so-called 'mermaids' of London society; freakish fakes wherein charlatans had procured monkeys' torsos and grafted them to the tails of fish. Had some Eastern mountebank sewn the bill of a duck on to the body of a mole or beaver?

He examined the base of the beak, surrounded as it was by a circular flap of somewhat leathery skin. Was the flap there to hide some surreptitious stitching? Try as he might, he could find no traces of thread that might prove the lie. Perplexed, he took a shallow dish and doused the entire pelt in water to see if he might loosen whatever glue was holding it all in place. It refused to fall apart. A horrible truth began to dawn on him. He was going to have to classify this monstrosity by naming its genus and species. But what was he looking at? A bird, a mammal, a fish or a reptile?

A WALK ON THE WILD SIDE

If there is communicable alien life out there in the galaxy, what is it like? The possibilities seem endless. Life on Earth is bewilderingly diverse. We find it everywhere, both microbial and multicellular, eating every conceivable type of food and inhabiting a myriad physical forms. Surely no one could have predicted the existence of the duck-billed platypus from first principles, no matter how much they knew about cellular biology, genetics, palaeoclimatology and the ecology of east Australian rivers? And if we can't second-guess life on our own planet, what can we ever hope to deduce about aliens?

Surprisingly, the situation isn't nearly as hopeless as you might think. As we are about to see, there's a wealth of information about life on Earth-like planets to be gleaned from life-as-we-know-it, and an abundance of informed speculation to draw on with regard to life-as-we-don't. Riddles such as the duck-billed platypus have given up their secrets, and the lessons we've learned are particularly instructive when it comes to alien-hunting. Not only are we about to meet a menagerie of exotic, exquisitely alien life forms inhabiting all manner of weird locations, but when it comes to rocky wet planets it turns out the Earth isn't the only game in town.

LOOKING IN THE LILY POND

As we know, the Kepler Space Telescope tells us that roughly one in five Sun-like stars has an Earth-sized planet in its habitable zone; that is to say, at the right distance to have liquid water on its surface. On average, that means the nearest such planet could be as little as twelve light years away.[1] According to Einstein, that means that even if we find it tomorrow, and invent speed-of-light space travel the day after, it's still going to take over a decade to get there. And if we fail to invent speed-of-light travel, with present technology the journey is going to take tens of thousands of years.

I tend to be optimistic about these kinds of things. Knowing a little about humankind, I can't help thinking that once we have found our nearest Eden we will want to get there as quickly as possible, and necessity, as they say, is the mother of invention.[2] But even if we conjure a technology that can drive our ships at, say, 10 per cent of the speed of light, the adventure of breathing the air on a planet similar to the Earth has to be several generations away at best.

So what do we do in the meantime? Are there any clues that we can glean from life on Earth as to what life on other Earths might be like? Happily, there are.

All life on our planet may share a common ancestor, but it has often inhabited parallel worlds. In some cases, these worlds have been evolving independently for tens of millions of years, with startling consequences. They are,

1 Tantalisingly, there is some evidence that Tau Ceti, the thirty-fifth most distant star from the Sun, may have an Earth-sized planet in its habitable zone. Tau Ceti is a single yellow dwarf, like our own Sun, and just so happens to be twelve light years away.

2 By then the father of invention may be global warming, but that's a topic for another book.

in short, the closest thing to setting foot on an Earth-like planet that we will experience in at least three generations. I'm talking, of course, about some of the greatest wonders of the globe: islands.

CONTINENTAL DRIFT

As we discussed in Chapter Six, at the time of the Permian extinction some 250 million years ago, all the present-day continents were joined together into one giant land mass, or supercontinent, known as Pangea. Thanks both to its size and a runaway greenhouse effect, the centre of Pangea became a dune-filled desert. We can still see its legacy today in the form of colossal deposits of sandstone; in fact, my own county of Cheshire in the UK is home to a sandstone ridge that is a relic of the Permian.[3]

The Permian, as you may know, marks the end of the Palaeozoic Era, when plants and animals colonised the land. The Mesozoic Era, which followed immediately after, saw Pangea begin to break up. The first big rift came towards the beginning of the Jurassic, when Pangea split into two: the northern lump, called Laurasia, was made up of North America, Greenland and Eurasia; while that in the south, called Gondwana, was made up of South America, Africa, Antarctica, India and Australia.[4]

By the beginning of the Cretaceous Period, roughly 140 million years ago, Gondwana was starting to break up. One by one, continents broke off the southern land mass and

3 See my handy guide to geological time on pages 190 and 191.

4 The name means 'Land of the Gonds'; the Gonds were an Indian tribe. The old name for Gondwana was 'Gondwanaland', meaning 'Land of the land of the Gonds'.

headed north to join Laurasia. Africa was the first to leave, splitting from South America to create the famous jigsaw puzzle by which the right hip of South America fits neatly into Africa's left. India followed soon after, making a beeline for Asia, while Africa headed north toward Europe. In the middle of the Cretaceous, about eighty million years ago, it was New Zealand's turn to take the plunge, while Madagascar split off from a still-migrating India.

Australia and South America were the last to leave, remaining joined to Antarctica until well into our present era, the Cenozoic. Australia departed first, in the Palaeogene Period, roughly forty-five million years ago. South America and Antarctica then hung on until the beginning of the Neogene Period, roughly twenty-three million years ago, at which point the Drake Passage opened up and the forests of Antarctica were replaced by permanent snow.

For the next twenty million years, South America migrated north, until eventually, at the beginning of our present Quaternary Period, an upwelling of volcanic rock created the Isthmus of Panama, and it became joined with North America. Finally, the globe had taken its present form, and a permanent ice cap began to form in the Arctic. The transformation from the tropical hothouse of the Jurassic, Cretaceous and Palaeogene into the icehouse of the Neogene and Quaternary was complete.[5]

---

5 While we're on it, a quick refresher on recent geological time periods. The Cenozoic Era spans the sixty-six million years since the impact that wiped out the dinosaurs at the end of the Mesozoic Era, and is subdivided into the Paleogene, Neogene and Quaternary Periods. The Paleogene was a hothouse, and the Neogene and Quaternary have been progressively parky. We divide the Paleogene into the Paleocene, Eocene and Oligocene Epochs, and the Neogene into the Miocene and Pliocene Epochs. The Quaternary is divided into the Pleistocene and Holocene Epochs. Roughly speaking, primates evolved at the beginning of the Paleocene, and hominins at the end of the Pliocene. Anatomically modern humans emerged in the Pleistocene, and human civilisation in the Holocene.

What all this means is that from the early Cretaceous onward, Gondwanan life effectively became marooned on a series of islands and continents, each of them like a tiny Earth. Fascinatingly, these alternate worlds paint a very different picture of how evolution might have turned out. On North America, for example, placental mammals came out on top, with marsupial mammals forced to extinction; on South America, however, the reverse was the case, and marsupial mammals dominated. In Australia, it was marsupial mammals that made the cut while the placentals perished; in New Zealand, no mammals survived at all.[6]

IF IT WALKS LIKE A DUCK … IT'S NOT NECESSARILY A DUCK

Madagascar is a case in point. Located in the Indian Ocean, just off the coast of East Africa, it is home to a fascinating array of mammals, none of which were present on the island when it rifted from India. As far as primates go it has no monkeys or apes, and no dogs or cats among its carnivores. All of the mammal species it does have – and it has a great many – are the descendants of creatures that either swam, floated or flew across from Africa and India.[7]

At some point round about sixty million years ago, a family of lemurs appears to have made the trip from East Africa. Not that they booked a package tour, of course; our best guess is that they were clinging to a tree during

6 The last known New Zealand land mammal died out in the Miocene around 16mya. Interestingly, several species of bat have repopulated New Zealand, where they have started to fill the vacant niche of ground-dwelling shrews.

7 I should point out that, like New Zealand, the fossil record of Madagascar shows that it was once home to native mammals that died out. The recently (2014) discovered 66 million year old fossil of *Vintana sertichi*, a strange-looking creature the size of a modern-day groundhog, is a case in point.

a hurricane when they got washed out to sea. A bunch of tenrecs – small, shrew-like mammals – followed in their wake roughly thirty million years ago, and other rafts bobbed up on the shore around twenty million years ago, carrying rodents and carnivorans, the mongoose-like ancestors of carnivorous mammals.

It's the lemurs that most people have heard of; present-day Madagascar is home to nearly one hundred different species, all descended from that first vagabond crew of early primates that crossed the Mozambique Channel on a fast-moving ocean current. The remarkable thing about lemurs is not only are they box-office catnip, as demonstrated by the fabulous *Madagascar* movies, but they are a striking visual echo of what the first primates must have looked like. Gaze into the eyes of a lemur, and you can almost imagine you are eyeballing one of our shrew-like ancestors in the tropical rainforest of the Palaeocene.

But while it's lemurs that get the glory, as far as our story goes it's the carnivorans that are the real stars. Do yourself a favour and Google a creature called a fossa. What pops up is something that looks like a cougar; so like a cougar, in fact, that many nineteenth-century taxonomists placed it in the cat family. The fossa has a head like a cat, a body like a cat and a tail like a cat. It climbs trees and can semi-retract its claws, just like a cat. Yet, intriguingly, a fossa isn't a cat. Instead, DNA studies have shown it to be a direct descendant of the African carnivorans that washed up in the Miocene some twenty million years ago.

Let's think about that for a moment. There have never been any jungle cats on Madagascar, yet the fossa looks just like one. How can that be? We'll never know for sure, but the smart money says that the fossa is a rather striking example of what we call convergent evolution. The same way of making a living – chasing small mammals through tropical trees – has produced two animals that are remarkably similar in appearance, despite the fact that they are only distantly related.

The fossa is one example, but Madagascar is home to many others. One of my favourites is the tenrec, several of which have evolved to look exactly like hedgehogs, despite being only distant relatives. Not that examples of convergence are limited to Madagascar; Australia, New Zealand and South America are all full of cases where unrelated creatures have ended up looking remarkably alike.

Where we find placental flying squirrels and moles in North America, for example, we find marsupial flying squirrels and moles in Australia, even though their last common ancestor was probably doing its best trying to avoid being trodden underfoot in the Cretaceous jungle. The sabre-toothed placental tiger of Ice Age North America had a much earlier doppelgänger in the sabre-toothed marsupial tiger of South America.[8] Even more striking, in my humble opinion, is the outward resemblance of the Australian thorny devil lizard, *Moloch horridus*, to the North American desert horned lizard, *Phrynosoma platyrhinos*. Both are desert-dwelling ant-eating lizards with blotchy markings and protected by exaggerated spines, and yet they are about as unrelated as it is possible for two lizards to be.

Such whole-animal convergences are striking, but they are only part of the story. Partial convergences are even more common, where creatures have very different body plans

8 The classic example of the placental sabre-toothed tiger is *Smilodon fatalis*, which first appeared in North America around 1.6 million years ago during the Pleistocene; along with the woolly mammoth and the dire wolf it belongs to an elite group of large mammals known as the Pleistocene megafauna. The marsupial tiger is the older *Thylacosmilus atrox*, which first appeared in South America in the late Miocene roughly eleven million years ago.

but similar body parts or behaviours. Flight, for example has evolved at least four times, in insects, pterosaurs, birds and bats, while, as we have already learned, camera eyes have evolved independently both in vertebrates and octopuses. The bearing of live young is estimated to have evolved over one hundred times among lizards and snakes, usually in response to cold climates, while hosts of animals we previously thought were related – flightless birds, for example – turn out to be only distant cousins.

RED IN TOOTH AND CLAW

It was just these kinds of partial convergences that so confused the European investigators of the duck-billed platypus. Following Shaw's examination in 1799, the classification of what is properly known as *Orinthorhynchus anatinus*[9] only became more problematic when, in 1802, the surgeon and anatomist Sir Everard Home reported that it possessed a cloaca, a single opening for the alimentary, reproductive and urinary tracts. That would seem to class the platypus among either amphibians, reptiles, or birds, and Home even theorised that it laid eggs. At this point the French anatomist Etienne Geoffroy Saint-Hilaire waded in, declaring that both the platypus and its countryman the echidna, or spiny anteater, represented an entirely new class of vertebrates. He named this new class the monotremes, from the Latin for 'single opening'.

9 Shaw gave it the name *Platypus anatinus*, with *Platypus* meaning flat-footed and *anatinus* meaning duck-like. Sadly the genus *Platypus* turned out to have been already taken by a wood-boring beetle. Meanwhile, the German naturalist Johann Friedrich Blumenbach had independently named the same creature *Orinthorhynchus paradoxcus* (bird-snout, paradoxical). Compromise was then reached with *Orinthorhynchus anatinus*.

Not that his fellow naturalists took much notice, preferring to join the mammal vs reptile fracas. The pendulum swung even further in the direction of reptiles with the 1823 discovery by the German anatomist Johann Friedrich Meckel that the platypus' spur was venomous. Venom, of course, is associated with reptiles like snakes and lizards. Three years later, however, Meckel published a paper that proved beyond doubt that the platypus had mammary glands, propelling the pendulum hard in the opposite direction. The entire platypus debate now hinged on one crucial question. Did it lay eggs?

The Aborigines were adamant that it did, but so rigid was the belief that mammals only gave birth to live young that, to begin with, few academics took the idea seriously. Even as late as 1884, the *Sydney Morning Herald* declared that any evidence to the affirmative must be 'examined and reported on by scientists in whom the world has faith, then all the scientific world will stand convinced and will believe where they have not seen'.

That same year, a tyro Scottish zoologist named William Hay Caldwell decided to blow his entire academic grant on travelling to Australia to settle the matter once and for all. An ecologically sensitive programme of shooting and dissecting platypuses had been in progress since 1834, overseen by the Australian naturalist George Bennett, the curator of the Australian Museum, Sydney, New South Wales, but the obnoxious Caldwell adopted what can only be described as a slash-and-burn approach.

In the Australian winter of 1884, assisted by an army of local Aborigines, he set up camp on the banks of the Burnett River in northern Queensland, and began slaughtering all the platypuses he could find. On 24 August, following a three-month period in which he had destroyed more than seventy platypus, he shot a female which had not only just laid an egg, but also had one in her uterus, ready to be laid. His triumphant telegram, 'monotremes oviparous, ovum meroblastic', was seen as the last word on the matter. Not only did the platypus lay eggs, it said, but those eggs were reptilian. The platypus was officially a conundrum.

THE EVOLUTION OF MAMMALS

If the study of islands and continents that have been isolated for tens of millions of years is the next best thing to finding an Earth-like planet, then to my mind at least the discovery of the duck-billed platypus is the next best thing to capturing an alien. And thanks to the Platypus Genome Project of 2008, we now have a much clearer idea of how this extraordinary creature fits into the grand sweep of evolution.

Part of the answer, as you might have guessed, is that the platypus is a descendant of a third group of mammals, the monotremes, which sits alongside the marsupials and placentals. The fossil record of monotremes is patchy, but suggests a radiation in the late Triassic or early Jurassic, eventually becoming extinct everywhere except in Australia, where they continue to thrive.

So far as we can tell from DNA studies, the last common ancestor of monotremes, marsupials and placentals probably lived some time during the Triassic, and appears to have been a furry egg-laying creature that suckled its young. Throughout hundreds of millions of years of evolution, the platypus has continued to lay eggs, while the other two surviving branches of the mammal family tree – marsupials and placental mammals – have instead evolved the ability to give birth to live young.[10]

So much for eggs. What about the platypus' other extraordinary traits, like its venomous spur and its duck bill? Fascinatingly, this is where convergent evolution comes in. Platypus venom is remarkably similar to reptile venom, but turns out to have evolved completely independently. The platypus' bill provides even more of a surprise. Everyone knew the platypus had to be doing something clever to be able to catch half its body weight in insect larvae at the bottom of muddy streams in the dead of night with its ears, eyes and nostrils closed; that something turned out to be what's known as electroreception.

Electroreception is very much de rigueur in fish, but almost unheard of in mammals. It turns out that the platypus' extraordinary bill, as well as mimicking shovellers such as the duck, is home to a vast array of receptors. As it swims, it swings its head from side to side, detecting the minute electric fields given off by its prey. In other words, despite having their last common ancestor way back in the Devonian, fish and platypus are both capable of sensing electric fields. Their evolutionary journeys couldn't have been more different, but given the same selection pressures – trying to make a living in murky water – the final destination was the same.

So what does all this tell us about aliens? Well, for starters it tells us that just because a planet is Earth-like, four billion years of evolution isn't necessarily going to produce anything remotely human. After all, on New Zealand and Australia

10 We used to say that the monotremes were therefore more 'primitive' than the marsupials and placentals, but, of course, that isn't the case at all. Monotremes have undergone just as much evolution as marsupials, for example; they continue to lay eggs because in Australia that's what works.

placental mammals didn't even make the cut, let alone the sub-division that we belong to, the primates.[11] And even if primates do evolve, and subsequently give rise to ground-dwelling, bipedal apes, there's no guarantee that they will fare any better. The Madagascan lemurs, for example, convergently evolved several 'ape' species, but they all went extinct.[12]

On the other hand, thanks to convergence, what comes around goes around. The islands and remote continents of the Earth show us that certain adaptations arise again and again: when it comes to life on planets like our own, we should expect to find the same notes, just not necessarily in the same order. Things like wings, eyes and teeth will be common on Earth-like planets throughout the galaxy, even though the creatures that possess them may be as unfamiliar to us as the duck-billed platypus was to nineteenth-century naturalists. These are, after all, the solutions that work time and time again, and towards which evolution will always stumble. And here's the rub. Intelligence, the vital commod-ity we need to find if we are to communicate with aliens, turns out to be a convergent trait.

FEATHERED APES

'I always wanted to fly,' says Nicky Clayton, as she spins her Audi TT around in a Cambridge side street, evading

---

11 The picture painted by phylogenetics shows placentals evolving first in Africa, and then spreading to Asia, North America, and eventually – once it joined via the Isthmus of Panama – South America.

12 The diversity of subfossil lemurs is astounding, and includes giant ape-like lemurs such as *Megaladapsis*, as well as orang-utan sized forms, and large ground-dwelling Gorilla-sized species. They also convergently hit upon 'sloths' (see *Archaeoindris* and others) with a wide array of sloth like subfossil lemurs and giant Aye-ayes.

yet another civic roadblock. 'That's why I love dance.' I nod enthusiastically. That explains why my bag is currently rolling around in the boot next to an entire wardrobe of lacy costumes and what can only be described as killer heels. Seconds later we are heading towards her laboratory in Madingley, and I briefly wonder whether the front of the car is going to achieve aerodynamic lift and grant her wish.

The building we arrive at has the understated hush of a local tennis club, with pavilion-style wooden huts surrounded by lawns and meshed enclosures. Instead of the thwock of brushed cotton on catgut, however, the air tinkles with birdsong. That's because, by night, Clayton is a dancer, but by day she is Professor of Comparative Cognition in the Department of Psychology at the University of Cambridge. What's more, she has made a considerable name for herself studying the intelligence of a previously overlooked species, the crow.[13]

Through a series of ingenious experiments, Clayton and her co-workers – her husband Nathan Emery is one of her collaborators – have demonstrated that, far from being 'bird brains', when it comes to intelligence, crows show a great deal of similarity to apes. Crows are foragers, for example, and love to hide food. In one famous experiment, Clayton and her team devised a sort of 'crow motel', where western scrub jays spent the night in one of two bedrooms. In one, breakfast was served in the morning; in the other, no such luck. After a stay of six nights, alternating between the two

13 I know what you're thinking: crows can't be that smart because they have tiny brains. Crucially, however, it's not size that matters; it's the ratio of size to body mass. A crow's brain may be the size of a walnut, but its body is extremely light, giving it a so-called encephalisation quotient, or EQ, on a par with apes. In fact, the western scrub jay, a particularly smart species of crow, has an EQ on a par with early hominins such as Australopithecus.

bedrooms, the jays were unexpectedly given nuts in the evening. Much as you or I might do, the crows hid food in the bedroom that came without breakfast, just in case.

That result is interesting, because it has been mirrored in similar experiments with apes. In one test, for example, chimpanzees and orang-utans were shown how to use a plastic hose to suck fruit juice from a container. Later, in a separate room, they were given a choice of four objects, one of which was the hose. Knowing that they would later encounter the container, the apes cunningly selected the hose. Apes and crows, in other words, aren't imprisoned in the here and now; they are capable of imagining their future and of planning for it.

Not only that, but Nicky's team have also demonstrated that crows are capable of designing tools, as well as reasoning, problem-solving, empathising and even deliberately deceiving one another. Other experimenters have shown that all of these abilities are shared by apes, though in my opinion, when it comes to tool-making, it's the crows that have the edge. And all this despite the fact that crows and apes have completely different kinds of brains, as you might expect with two species whose last common ancestor was an amniote[14] that roamed the Carboniferous forests some 300 million years ago.

So what caused the intelligence of crows and apes to converge? Clayton makes some intriguing suggestions. Both animals are highly social, and, as we all know, to get ahead in society you need an ability to play politics. That requires brainpower, and it may be that group living is one of the drivers towards animal smarts. Second, both apes

14 Meaning something that was capable of laying hard-shelled eggs on land. Amniotes then split into the synapsids and sauropsids. Synapsids eventually gave rise to mammals and sauropsids to reptiles and birds.

and crows are foragers, living off seasonal foods that are tricky to identify, hard to get at and distributed over a wide area. In this case, it's not just the early bird that catches the worm, but the one that's gifted enough to drop stones into a pitcher of water.[15]

And third – and to my mind, most significantly – both chimpanzees and crows first appeared between ten and five million years ago, a time of rapid climate variation as the Earth limbered up for the present Ice Age. Interestingly, that's also when our own hominin line diverged from the common ancestor it shares with *Pan*, the genus to which chimpanzees and bonobos belong. One way to get around a changing climate, of course, is to use your intelligence to help you find food and shelter. Could what Clayton calls the 'clever club' of apes, crows, parrots, dolphins and elephants be a direct result of unpredictable weather?

For those of us seeking to contact aliens, the implications of all this are profound. Just as we can expect the inhabitants of Earth-like planets to fly, bite and see by the light of their home star, we should also expect them to be smart. Because despite what we might want to believe, intelligence is not a uniquely human trait; we share it with crows, dolphins, elephants, and no doubt a whole zoo of other terrestrial species yet to be investigated.[16] Its basic components – things like reasoning, problem-solving, imagination, memory and mental time travel – crop up time and again.

---

15 I know that sounds like I'm culling my data from Aesop's Fables, but Sarah Jelbert at the University of Auckland actually did this experiment. In fact, if you put 'crow and pitcher' into *New Scientist's* YouTube channel you can watch a version for yourself.

16 Just for starters, and in no particular order: octopuses, dogs, cats, rats, whales, parrots and pigs. The problem-solving abilities and tool use of octopuses is particularly significant, because they are about as far removed in animal evolution from the other examples as it is possible to be.

WAR OF THE WORLDS

So we have half an answer to the question of what an intelligent alien might look like. First, it will have a complex brain. Whether that brain is centralised, as it is in birds, mammals and dolphins, or decentralised as in octopuses, we can be less sure. It will have acute senses, whether attuned to light, sound, heat, chemicals or electric fields. It will most likely eat a wide variety of foods, be extremely dexterous with physical objects, communicative and social. Large portions of its long life will be spent learning from its society, parents and peers. And it will quite possibly have evolved in an ever-changing terrain, where it used its intelligence to stay one step ahead of the game.

As to whether it has six legs, fur, feathers or a two-inch-thick coating of mucus, all bets are off. Certainly, if it resembles any of the family groupings we recognise on Earth – vertebrates, or invertebrates, for example – that will be a fluke. There is nothing special about the order in which species have diverged here on Earth; if we re-ran evolution it might just as well be a ray-finned fish rather than a lobe-finned fish that gave rise to the first amphibians, for example, and we might have inherited six digits per hand rather than five.

If it comes to that, of course, if we re-ran Earth evolution we might not even get as far as the lobe-finned fishes. There are considerable hurdles to jump on the track to complex life as we know it, with one of the hardest to clear being the creation of the eukaryotic cell and its energy-giving mitochondria. Another is the evolution of oxygenic photosynthesis, whereby chlorophyll is used to harness light energy to rip electrons from water and stuff them on to carbon dioxide to make sugars. Fail to make it past either of those and all you get is an ocean full of bacteria, much as we had for the first two billion years of life on Earth.

As we shall see in a moment, there are reasonable arguments as to why the eukaryotic cell might be convergent, and how we might get by with ordinary photosynthesis rather than the souped-up, oxygenic kind, but however you slice it, the sixties *Star Trek* trope of landing on an opulently flowered Eden inhabited by nubile blondes and beaux with negligible body hair is very much an outside bet. Even though it ticks all the above boxes for brain size, dexterity and communication, the intelligent alien on the other end of the phone might look more like a crab, or a spider, or an octopus; even more likely, it will resemble none of the above, but for a sucker here and an eye there.

Of course, it takes a lot more than just intelligence to make a communicable alien. Humans are smart, but not that smart. What really sets us apart is civilisation; what made that possible was the invention of farming. As we entered the Holocene, and the snows retreated, everyone got their hoes out. Or to be precise, they got the halters out; the hoes came shortly after. Farming, too, appears to be highly convergent, emerging roughly 11,500 years ago in the stable climate of the Holocene in South East Asia, the Levant, the Fertile Crescent, South America and Europe. Oh yes, and roughly fifty million years ago in the Attine ants of the Amazon rainforest.

ALIEN ANT INVASION

Not that they were farming rice and rye, of course; the preferred crop of the Attines is fungus. One particularly sophisticated group, the leafcutter ants, live in civilisations over five million strong, with precise divisions of labour.

Their underground nests reach enormous size; one excavated in Brazil in 2012 was fifty square metres across and eight metres deep. A network of tunnels provide access and ventilation to large underground chambers where the fungus is fed with leaf mulch; the fungus breaks down the cellulose in the leaves so that both it, the ants and their larvae can digest it. As the fungus grows, the ants prune it, fertilise it and even treat it with antibiotics when it becomes infected with other parasitic fungi.

I'm sure by now you are getting the picture: there is almost nothing about humans that is unique. There are plenty of other creatures out there that walk on two legs, give birth to live young, or farm for a living. Ants, of course, aren't great shakes in the cognitive department, which is why they aren't planning a fly-by of Europa; not on this planet, anyway. On the one hand, that's a little disappointing; we humans love to feel important, and it's a bit disconcerting to discover that a six-legged creature in the Amazon jungle mastered the antibiotic before we did, but when it comes to our quest to find other communicable life in the galaxy it should give us great heart.

All the important stuff like intelligence, language, tool use, farming and civilisation is convergent; it has evolved before, in countless other species. And while no one was waiting for humans to be the ones that got their act together first, by chance that is exactly what happened. True, evolution has not been conducted with us in mind; when the first archaeon left the vent, it wasn't with the express intention of one day wandering about Soho with a copy of *Time Out*.[17] But the ratchet of natural selection has spent the last four billion years layering complexity

17 *Private Eye*, maybe, but not *Time Out*.

upon complexity; it was just a matter of time before one species lucked out.[18]

THE INFORMATION SUPERHIGHWAY

And the precise way in which we lucked out, of course, was that we came up with a method for storing information outside our physical bodies. As previously mentioned, the first phonetic writing system is believed to have emerged in Sumer in Mesopotamia around 3100 BC. As Claude Shannon would have noted, this marks a profound change. Before writing, DNA was the only way nature had of making a 'hard copy'; afterwards, anything anyone had a mind to write down could be preserved for generations.

From Mesopotamia, the technology of writing spread quickly west to Ancient Egypt, and then around the Mediterranean, courtesy of the Phoenicians. The Ancient Greeks adopted the Phoenician alphabet, and bequeathed it to the Romans. We have no way of knowing whether there were cavemen as wise as Socrates, but thanks to writing, even 2,400 years later we are intimately acquainted with his every thought.

Information, let's not forget, is a physical thing. While the Sumerians had scored their marks in wet clay, the Egyptians had something far handier – papyrus – from which they were able to make the first books and eventually to found

18 That's not to say every species on Earth has become more complex, obviously. The vast majority of life on the planet is still single-celled, in the form of archaea and bacteria. Organisms can lose complexity as well as gain it, or even stay much the same. Like many cave-dwelling species, the Hawaiian Kaua'i cave wolf spider has lost its eyes, for example, and the coelacanth is doing a pretty good impression of other coelacanths that lived 400 million years ago. It's average complexity that increases over time, which is why we would be so surprised to find a fossilised ichthyosaur with a brain-to-body ratio bigger than a bottlenose dolphin, or an Attine ant farm with lifts and designated parking spaces.

the first library in roughly 300 BC in Alexandria. Writing not only meant that information could be preserved; it could also be copied. Reportedly, when a ship cast anchor in the port of Alexandria, any written works on board would be confiscated on behalf of the library and judiciously copied.

As seekers of alien civilisations, we would do well to note that besides the Mediterranean and Middle East, writing appears to have independently evolved in both South America and China, so we can be reasonably sure that it, too, is convergent. The Chinese, in fact, not only invented modern paper, but also the first woodblock printing. They were also the first to invent movable type,[19] but for some reason – the vast number of characters in Chinese script, perhaps – the idea appears not to have caught on.

In AD 751, the western expansion of the Tang Dynasty was halted by the Abbasid Caliphate at the Battle of Talas, and the Chinese prisoners of war revealed the secrets of papermaking to their captors. By AD 794, paper was being manufactured in Baghdad. Throughout the Middle Ages, Islamic scholars built upon the classic Ancient Greek texts, laying the foundations of modern science and mathematics.[20] Thanks to the Crusades, Muslim scholarship and papermaking technology eventually began to reach Western Europe, and by the late thirteenth century first Italy, and then France and Germany, became centres for papermaking.

Another leap then came with the invention of the printing press in Gutenberg in 1450; as information flooded across Europe, it suddenly switched from being a backward region

19 In 1041 a Chinese alchemist named Pi Sheng invented movable clay type; and in 1313 a Chinese magistrate named Wang Chen invented movable woodblock type.

20 It is the Islamic mathematician al-Khwarizmi (c. 780–850) who gives us the word 'algebra', being the Anglicised form of 'al-jabr', one of the mathematical operations he used to solve quadratic equations.

to being the proud home of scientific greats such as Leibniz, Kepler and Newton. Their discoveries in turn ignited the industrial revolution of the mid-nineteenth century with its mechanisation of farming, manufacturing and the production of energy. Finally, the present digital age has seen every bit of information in existence made available online, orbiting the planet in an impenetrable electromagnetic swarm.

As humans, we find it difficult to see our technology as part of nature, as much the product of natural selection as, say, the anthill or the shell of a snail. Our snaking motorways and dotted radio masts appear to us to be a very different thing from the veins of a leaf, or nodes of a spider's web, but of course they are fundamentally the same. To see something like the internet as a natural extension of the human organism feels absurd, but it's important to ask the question: what is all this information for?

Take the evolutionary biologist's view, and the answer comes back with resounding clarity: it's not 'for' anything. Pieces of information that are good at getting themselves copied will eventually come to dominate the world, whether they are the gene for blue eyes or the code for a cute cat video. One fruitful strategy is to improve the fitness of their host. The information in your iPad improves the chances that you will have offspring, just as that in your DNA does. Admittedly, it's a bit hard to see the direct effect that reading a good Dick Francis novel has on your capacity to bear children,[21] but, believe me, that may well be why he's there. Zoom out and look at the big picture and the facts are undeniable. Since the invention of writing, human population has grown with exponential fury.[22]

---

21 It keeps you sane in a crazy world, thereby enabling you to avoid traffic and eventually go to bed with someone.

22 Human population is estimated to have been less than ten million at the beginning of the Holocene. In 1804 it reached one billion, and two billion in 1925.

Throughout this book we have tried to figure out what intelligent life on Earth can tell us about intelligent life on other planets. The first vital step was the creation of the cell; the next was photosynthesis. The evolution of the eukaryote enabled the leap to multicellular life; that was followed by the evolution of animals, or metazoans. Finally, the shift from DNA to silicon as a means of storing information enabled the accomplishments of human civilisation to transcend anything that might be possible for an individual human. Writing is convergent, and with writing comes an explosion of information that transforms a tribe into an accomplished technological civilisation. If there are other Earths out there, there could well be aliens like us to talk to.

But what about planets that aren't like the Earth? Might we find intelligent communicable life on them? And, if we do, what would those aliens look like? It's time to cast our net a little wider, and fish our neighbouring solar systems for life-as-we-know-it. And, finally, we shall need to cast our net a little wider still, and fish for the ultimate prize; life-as-we-don't. Could there be intelligent beings out there whose chemistry is based on an element other than carbon? Does life even need chemistry at all? Could it take root in a dust cloud, or on the surface of a star? And how would we know it if it did?

SUPER-SIZE ME

If we've learned one thing in the course of our journey together, it's that here on Earth the vital steps that led to the emergence of communicable life took a great deal of time, a total of roughly four billion years.[23] The problem is that while the lifetime of Sun-like stars is something like ten billion years, after half that time they start to run out of hydrogen and heat up. Bit by bit, over the next 500 million years the Sun will vaporise the world's oceans and turn our planet into a smog-choked desert. Put simply, the window of opportunity for stars like the Sun to develop creatures like ourselves is a little bit on the snug side.

Fine, you might think: let's find the longest-lived stars out there and look for Earth-sized planets in their habitable zones. But there's a problem. The longest-lived stars are the smallest, dimmest ones, known as red dwarves. Red dwarves have lifetimes of up to trillions of years, but not only are they far less stable than the Sun, spitting out all sorts of nasty radiation, their habitable zones sit much closer in. That means any wet rocky planets they happen to have will be right in the firing line.[24]

Happily for us, however, there's an alternative. There's an intermediate size of star, the so-called orange dwarves,[25] that are perfect for nurturing life. Not only do they have

23 In broad strokes, cellular life emerged from the vent around four billion years ago, evolved oxygenic photosynthesis around three billion years ago, evolved the eukaryotic cell around two billion years ago, multicellularity around 1.5 billion years ago, then complex multicellularity around 575 million years ago. Plants moved on to land in the Ordovician roughly 480 million years ago, amphibians followed them in the Devonian 375 million years ago and 2009 saw the release of the alpha version of Minecraft.

24 There might be a way round this. Since red dwarves are relatively lower in mass, a rocky planet in the habitable zone may well be in tidal lock with the parent star. If one hemisphere is always facing away from the star it would be shielded from radiation doom, and could have energy transferred to its radiation-shielded night side via the atmosphere. Alternatively, a giant planet might generate a large protective magnetic field that could shield itself or its moons from the worst effects.

25 It was a while ago, so here's a brief reminder that we label the hydrogen-burning stars as follows, brightest to dimmest; O, B, A, F (yellow/white dwarf), G (yellow dwarf), K (orange dwarf) and M (red dwarf).

long lifetimes, in excess of fifteen billion years, but they are stable, like the Sun. In fact, they even emit much less harmful ultra-violet light than Sun-like stars; here on Earth we only managed to get decent protection from UV after the formation of the ozone layer, following the Great Oxidation Event roughly 2.3 billion years ago. Perfect. So we should look for Earth-sized planets in their habitable zones, right? Wrong.

Unfortunately, when it comes to long lifetimes, small rocky planets like the Earth won't cut it. One of the most crucial ingredients for carbon-based life, as we know, is plate tectonics. Not only does outgassing from volcanoes keep the atmosphere full of carbon dioxide, and therefore keep geologically active planets from freezing over, but, according to our pet theory, it's in volcanic vents on the sea floor that life gets its start. No volcanoes, no life. And to be volcanic, of course, the centre of a planet needs to be hot. The problem is, small rocky planets like the Earth will cool down long before an orange dwarf burns out.

This is where super-Earths come in. Being several times more massive than the Earth, they keep their heat longer and can easily remain volcanically active for the long lifetime of an orange dwarf star. Along with a molten core, of course, comes a magnetic field, and protection from harmful cosmic rays and solar flares. Some, such as René Heller of the Institute of Astrophysics, Göttingen, even go so far as to call such planets 'super-habitable', better than Earth for the evolution of life. And the kicker is that, unlike diminutive Earth-sized planets, they will show up well in the next generation of telescopes.

INTELLIGENT LIFE 2.0

So what would life on a super-Earth in the habitable zone of an orange dwarf star look like? Just as in the case of an Earth-sized planet, while it is impossible to predict the precise route that evolution will take to get there we can make some informed guesses as to where it will end up. And thanks to the fact that it can run for nearly twice as long as it can here on Earth, that may be a very interesting place indeed.

A CLOCKWORK ORANGE

Let's start with what physics can tell us about a larger-than-life Earth planet. No prizes for guessing that a bigger Earth means stronger gravity, but what effect will that have on the land, oceans and atmosphere? Interestingly, quite a lot. Calculations show that plate tectonics kicks off when planets are roughly Earth-sized, and switches off when they reach around five times Earth mass. At the lower bound, you get deep oceans and large continents; by the time you reach twice the Earth's mass, stronger gravity produces shallow oceans with island archipelagos.

That's exciting, because shallow oceans and islands, as we know, are great for biodiversity. Even today, we find the interiors of large continents and the depths of beetling oceans more sparsely populated, while islands and lagoons teem with life. But we are getting ahead of ourselves. Would we expect life on a super-Earth to follow the same path it did here on Earth, with single-celled life emerging in the oceans, evolving complexity, then invading the land?

Again, in my view, the path may be different, but the destination remains the same. According to our pet hypothesis, single-celled life started in the vent at least twice: once as bacteria and a second time as archaea. Anything that happened more than once is a prime candidate for convergence, so we can assume that single-celled life is a given. Both bacteria and archaea then navigated the next step; namely, they weaned themselves off the electrical energy of the vent, and learned to harness light energy, also known as photosynthesis.

Of course, this early photosynthesis involved using light energy to rip electrons from chemicals that surrounded the vent: hydrogen sulphide, for example, was a favourite; iron was another. Those electrons were then forced on to molecules of carbon dioxide, creating sugars. Eventually, in one type of cell only, the cyanobacteria, it gave way to a much more complicated, but altogether more successful process: oxygenic photosynthesis. Here, light energy was used to rip electrons from water, also creating sugars, but this time releasing oxygen as a waste product.

Primitive photosynthesis, in other words, emerged several times; oxygenic photosynthesis only once. To my mind, that means we have to surrender to the possibility that other Earths and super-Earths will not be flooded with highly combustible oxygen. But that needn't cramp our style. After all, free oxygen was a huge problem for primitive life. Not only was it toxic, but it oxidised atmospheric methane, removing a vital greenhouse gas from the atmosphere and triggering a Snowball Earth in the form of the Huronian glaciation.

Eventually, of course, life adapted, and eukaryotes which could burn oxygen in their mitochondria came to rule the Earth. But couldn't eukaryotes have evolved anyway, without the oxygen problem? Endosymbiosis, which created the eukaryotic cell, has happened more than once; to take a notable example, just as we think an archaeon engulfed and enslaved a bacterium to produce the eukaryotic cell, so we believe that a eukaryotic cell engulfed a cyanobacterium to produce the plant cell.

The eukaryotic cell is the McLaren P1 supercar to the prokaryote's Mazda MX-5; once it appears on the scene, the brakes to complex multicellularity are well and truly off. Here on Earth, complexity has arisen in animals, fungi plants and algae, and it's a fair bet it will emerge on our super-Earth, too.[26] Which complex forms will first populate the oceans, however, is anyone's guess. The strange creatures of the Ediacaran are probably our best rule of thumb, but who knows how things will progress from there?

All we can rely on, really, is the same extraordinary adaptive power that we find here on Earth. If some single-celled organism does manage to crack oxygenic photosynthesis, no doubt plant-like forms will invade the land just as they did here. They might not be green, of course. No one is entirely sure why chlorophyll absorbs mostly blue and red light: it may just have been the nearest pigment to hand, or it may be optimised in some way to the kinds of light photons that make it through the atmosphere. Whatever the answer, there's no reason that alien chlorophylls would have to be green. To our eyes, the plants on an orange dwarf planet could just as easily appear yellow, or even black.

And if oxygenic photosynthesis fails to emerge, well, no doubt life will find another way. Complex multicellular organisms that use sunlight to synthesise sugars – a weird cross between a plant and a fungus, perhaps, or a

---

26 Of course, in terms of complexity, plants are streets ahead of algae and fungi, and animals are lightyears ahead of plants.

bioluminescent slime mould – will eventually populate the oceans and then the land. Once they do, other complex creatures that are capable of eating them will follow.

Again, none of their body plans would necessarily be recognisable, but weapons, armour, jointed limbs, eyes, mouths and brains have evolved so many times here on Earth it's reasonable to assume they will put in a regular appearance. Maybe the thicker atmosphere will encourage flight, and the air will swarm with all manner of airborne creatures; maybe the many island habitats will produce a greater diversity of terrestrial life than here on Earth. We can't be sure. Eventually, however, one or more intelligent species will develop civilisation, and that's when the real fun will start.

BACK TO THE FUTURE

The possibility that super-Earths are even more habitable than our own planet brings one thing into sharp focus: any alien civilisations we do find are likely to be much older than our own. That makes it hard to know what to look for. It's hard enough to imagine where human civilisation will be in five years, let alone an alien civilisation in 500 million. How would such planets show up in our telescopes?

Kepler 62 is a perfect example. Discovered on 18 April 2013, it's an orange dwarf star roughly 1,200 light years away in the constellation of Lyra. Not only might it be as much as eleven billion years old, but it also has two super-Earths in its habitable zone. Assuming that the Earth is average, and it takes four billion years to raise a technological civilisation, that means intelligent communicable alien life could have existed on either or both of those planets for a billion years.[27] Where do we start? What do we look for?

We certainly have no way of predicting what one-billion-year-old technology would look like. But given what we know about life here on Earth, we can still make a shopping list of things to look out for. From the Second Law of Thermodynamics we know that all life, no matter how intelligent, dissipates energy as heat. Heat, as you know, is essentially infra-red light. So all we have to do is point one of our infra-red space telescopes at a nearby orange dwarf with super-Earths in its habitable zone and see if they are giving off a splurge of infra-red light, right?

Sadly not. Even though orange dwarves are far dimmer than our own Sun, they are still way too bright for even the next generation of infra-red telescopes such as the James Webb to be able to pick out the heat signature of a civilisation. We might have more luck picking out what's known as a Dyson sphere – a civilisation which has blacked out its home star with light-absorbing satellites – but there are lots of darkish things in the sky which emit in the infra-red, such as protostars surrounded by dust, making their existence very hard to prove.[28]

As I hinted earlier, however, it might be within the capabilities of the James Webb, and indeed the forthcoming ground-based European Extremely Large Telescope (E-ELT)[29] to pick up enough infra-red from transiting super-Earths to be able to work out what they have in their

27 Not 6.5 billion years, remember, because of GRBs. Anything older than five billion years will be toast.

28 Nevertheless, Fermilab run a programme on the IRAS infra-red satellite which attempts to do just this. So far they have identified seventeen 'ambiguous' candidates, four of which they class as 'amusing but still questionable'.

29 Planned for 2022.

atmospheres. We could look for the distinctive spectra of oxygen and methane, for example, both of which we know are produced by life here on Earth, or pollutants such as chlorofluorocarbons (CFCs). But to really examine the infra-red given off by nearby super-Earths we are going to need something game-changing.

And that something might just be NASA's New Worlds mission. Currently at the drawing-board stage, New Worlds is an ingenious way to solve the problem of life-bearing planets being very dim compared to their home stars. Essentially, a space telescope is launched together with an occulter, a colossal starshade which can be manoeuvred to block out the light of the home star so that the planets show up. That way we can get a clear look at both the infra-red and the visible light that such planets give off, and maybe even resolve surface features like oceans and land masses. Imagine the excitement if the images show networks of light, just as you might see on the dark side of Earth.

CLOSER TO HOME

So much for Earths and super-Earths. What other types of planet might harbour life? The first place to look, obviously, is our own solar system, where currently there are two leading candidates: Europa and Titan. Europa, you'll remember, is one of the four moons of Jupiter picked out by Galileo in 1610, photographed by the *Pioneer* and *Voyager* probes in the 1970s, and last visited by NASA's *Galileo* spacecraft in 2003.[30]

---

30 The *Galileo* spacecraft was launched in 1989, arrived in 1995, and spent eight years in the Jovian system. It was deliberately crashed into Jupiter to avoid contaminating its moons with Earth life.

Crucially, the Galileo mission discovered that Europa is an icehouse world, with a warm salt-water ocean contained within a crust of ice. There are clays sitting on top of the ice, suggesting a recent collision with an asteroid or comet; as we know, comets and asteroids also carry organic materials, so there's a real chance Europa's ocean is seeded with long-chain carbon molecules. Not only that, but the Hubble Space Telescope has recently spotted plumes of water vapour billowing out from Europa's south pole, suggesting that we might not have to dig through the ice to sample its ocean.

There are two missions slated to launch to Europa in the early 2020s. The ESA's Jupiter Icy Moon Explorer (*JUICE*) will arrive around 2030, though NASA's as-yet-untitled Europa mission might get there first if it manages to piggyback their new Space Launch System. As the name suggests, *JUICE* will also fly by Ganymede and Callisto; the NASA mission, on the other hand, will focus specifically on Europa. By the time they're done, we'll know exactly what temperature the ocean is, how deep and how salty. Even more excitingly, the NASA mission may yet attempt to fly through one of Europa's water plumes with its mouth open to look for interesting chemistry.

If it's looking positive, no doubt a landing mission will soon follow; anything from a probe that touches down near the south pole to sample surface water from the plume, to a submarine capable of burrowing down through the ice into Europa's ocean. What might the life forms down there look like? The short answer is we have no idea. Assuming we don't come face to face with the Europan equivalent of the Kraken, the smart money would be on some form of microbial life. Beyond that, we would be looking for complex carbon molecules – things like amino acids and sugars – but, of course, Europan life might use a completely different suite of carbon molecules to the ones we find here on Earth. Even if we find it, we might not immediately recognise it.

Titan is another enticing prospect, although a new mission to follow up on the ESA's *Huygens* probe is still several decades away. Titan, you'll remember, is truly an alien world. Its climate and terrain are remarkably similar to our own, except methane takes the place of water, and water ice takes the place of rock. Methane clouds float through its thick nitrogen atmosphere, showering methane rain into lakes and dumping methane snow in its mountains.

Here, the most exciting mission would be one where a submersible is landed on one of Titan's methane oceans, and explores the depths for signs of life. Could its methane lakes provide the perfect solvent for silicon chemistry? Silicon belongs to the same chemical group as carbon, and is also capable of forming long-chain molecules. Might it be possible that complex life could evolve on Titan, only based on silicon rather than carbon?

The astronomer Maggie Aderin-Pocock thinks so; in a 2014 article she proposed that such aliens might take the form of enormous floating jellyfish, buoyed by enormous bladders, sucking in atmospheric nutrients through giant mouths, and communicating with one another via pulses of light. Oh yes – and their bottoms would be orange to camouflage them in the hazy Titanian skies. But if you think that's weird, you need to know about Vadim Tsytovich and his living clouds of dust.

SEARCHING FOR THE CORKSCREW

Might clouds of interstellar dust be alive and have seeded the first life on Earth? That's the intriguing proposition of the veteran Russian plasma physicist Vadim N. Tsytovich.

In 2007 he published a speculative paper in the *New Journal of Physics* in which he described how plasmas – that's clouds of charged particles to you and me – might actually organise dust grains into self-replicating helical structures that tick every box for living things.

Plasma, as you may know, is the fourth state of matter after solid, liquid and gas; essentially, it's what you get when gas molecules break down to form charged particles, or ions. A classic example would be a neon sign, a glass tube of neon gas that becomes a light-emitting plasma when it has electricity passed through it. Another would be an electric spark, where free electrons from cosmic rays or background radiation are accelerated by an electric field, and collide with air molecules causing them to form a plasma. Electricity then flows through the plasma producing light, heat and sound.

Although we don't tend to come across them that often on Earth, plasmas are actually one of the most abundant forms of matter in the universe. Not only are there vast filaments of plasma out in intergalactic space, but we also find them in interstellar molecular clouds, in the proto-planetary disks that surround young stars and the upper atmosphere. In all these cases they are mixed with dust, which they organise into fascinating structures known as 'plasma crystals'.

In modelling such crystals, Tsytovich discovered that they have some remarkably lifelike properties. Essentially, under the right conditions they are capable of forming double-helical structures that are highly reminiscent of DNA. Not only does the width and length of the helices change, providing a way of encoding information, but in some simulations they would divide into two, effectively reproducing themselves.

We have yet to create such structures in the laboratory, but Tsytovich is convinced that the helical dust structures within plasmas exhibit all the properties of living matter. After all, they feed on the energy of the plasma, they reproduce and they evolve into permanent complex structures. Could they in some sense be alive? And, for that matter, where exactly is the boundary between living and non-living things? As Fred Hoyle suggested in his 1957 novel *The Black Cloud*, could there be intelligent beings made up of nothing more than electrically charged dust?

COOKING UP SOME ANSWERS

Last night I was preparing some pasta for my middle son. I took a small pan of boiling water, gave it a glug of olive oil and chucked in a handful of pasta tubes. Ten minutes later I lifted the lid to check on it. All the tubes were standing up on end, crowded cheek-by-jowl on one side of the pan, surrounded by boiling water. From a jumbled mess in the bottom of the pan, they had been jostled into a highly ordered state.[31] Why should that be?

We see this sort of thing all the time in nature, but as yet we don't quite have the physics to describe it. Our thermodynamics – our theory of how energy and information are related – has really only been figured out for systems that are in balance, or, as a physicist would say, equilibrium. But most systems in the real world aren't like that. Supply a thing with energy, and more often than not it organises itself. Thump a tank of water with a regular beat and you will produce a pattern of ripples. They may look pretty, but their deeper purpose is to dissipate the energy of your thumps as quickly as possible. In the same way, when I heat

31 The olive oil may be the crucial ingredient here. Omit it – and fail to stir – and your penne will limpet to the bottom of the pan in a congealed mess.

the bottom of the pan, the pasta organises itself because that way it can dissipate the heat more rapidly into the room than if it were jumbled up.

There is some very interesting work being done in this area,[32] and it indicates that something very similar is going on with living systems. The gravitational energy of the Sun needs dissipating. The Sun becomes layered, and initiates nuclear fusion, radiating highly organised light. That light then enters the biosphere, where carbon molecules become ordered into life in order to dissipate the energy of that light. Life exists to speed the heat death of the universe.

Our problem as organisms, of course, is that we don't see the bigger picture. We see our existence as a battle; an attempt to maintain order in a world that demands chaos. But we are missing the point. In our struggle to stay alive, we are serving the universe because we are dissipating energy. The universe doesn't want those fossil fuels in the ground, it wants them burned. What better way to do that than to throw up an intelligent civilisation with an insatiable thirst for energy?

Following that logic, intelligent life should truly be everywhere. Wherever there is an energy source we should expect to find that matter organises around it, helping it to dissipate its energy as fast as possible. In the case of a star, one way is to form a solar system. Within the gas, dust and planets of that solar system, matter will organise itself into spheres, weather patterns and life. The reason we developed intelligence and technology is the same one that caused bacteria to develop photosynthesis; it's a great way to dissipate the energy of the solar system.

---

32 I'm thinking specifically of Jeremy England of MIT and his 2013 paper 'Statistical physics of self-replication', available from all good search engines.

There is nothing fundamental that separates life from non-life. It is all just matter. We are ripples. Seen this way, the aliens are everywhere. Some are no more than the smooth regular pebbles on the beach, or the bubbles of carbon dioxide in your fizzy drink. Still others are moulds and fungi. Others are quantity surveyors. And others are helical crystals in clouds of dust.

CHAPTER EIGHT

Messages

In which the author quarries for a Rosetta Stone, and blasts the skies with Big Data.

> *With what meditations did Bloom accompany his demonstration to his companion of various constellations?*
>
> *Meditations of evolution increasingly vaster: of the moon invisible in incipient lunation, approaching perigee: of the infinite lattiginous scintillating uncondensed milky way, discernible by daylight by an observer placed at the lower end of a cylindrical vertical shaft 5000 ft deep sunk from the surface towards the centre of the earth: of Sirius (alpha in Canis Major) 10 lightyears (57,000,000,000,000 miles) distant and in volume 900 times the dimension of our planet: of Arcturus: of the precession of equinoxes: of Orion with belt and sextuple sun theta and nebula in which 100 of our solar systems could be contained: of moribund and of nascent new stars such as Nova in 1901: of our system plunging towards the constellation of Hercules: of the parallax or parallactic drift*

> *of socalled fixed stars, in reality evermoving wanderers*
> *from immeasurably remote eons to infinitely remote futures*
> *in comparison with which the years, threescore and ten, of*
> *allotted human life formed a parenthesis of infinitesimal*
> *brevity.*
>
> JAMES JOYCE, *Ulysses*

On 20 July 2015, at the Royal Society in London, the Russian internet billionaire Yuri Milner called a press conference. Accompanied by luminaries such as Stephen Hawking, Frank Drake and Martin Rees, as well as by Carl Sagan's widow and co-collaborator on Sounds of Earth, Ann Druyan, Milner announced a game-changing new SETI initiative called *Breakthrough Listen*. Over the next decade, mankind will search the nearest million stars and the nearest hundred galaxies for signals. If ET is calling, we are about to pick up the phone.

In essence, what Milner has done is provide SETI with a big pot of cash[1] to buy time on three of the world's most powerful telescopes. The Green Bank Telescope in West Virginia and the Parkes Radio Telescope in New South Wales will search for radio signals,[2] while the Lick Telescope in California will hunt for optical laser transmissions. Up until now, it has been very hard for SETI to afford more than one day a year on this kind of kit, which is just one of the reasons why they built the Allen array. Now they will be getting thousands of hours a year, vastly increasing the speed, scope and resolution of their search.

SETI has been conducting searches for visible light

1 $100 million.

2 Green Bank is, of course, the telescope on which Frank Drake conducted the first ever SETI search, when, in 1961, he pointed it at our two nearest Sun-like stars Tau Ceti and Epsilon Eridani.

transmissions – known in the trade as 'optical SETI' – for over a decade, but the recent step increase in our own use of lasers makes the *Breakthrough Listen* search seem all the more timely. In January 2014, NASA used a laser to broadcast an image of the *Mona Lisa* to the Lunar Reconnaissance Orbiter (LRO), a robotic craft which is currently making a survey for future Moon landings. That was swiftly followed by the International Space Station's OPtical PAyload for Lasercomm Science (OPALS), where NASA proved they were able to send data much faster over laser than they are currently able to with radio waves. Could it be that in looking for radio transmissions we are way behind the times? Maybe ET has moved on, and communicates with its satellites and lunar base stations using lasers?

Adding the Lick Observatory to the search means we can ramp it up for just this kind of signal. One of the disadvantages of radio waves is that they spread out over a wide area, dissipating their power; lasers, on the other hand, are much more directional. The snag is that to detect a laser beam you need to be close to its line of sight. There's also the risk that laser signals would be heavily encrypted; after all, anyone broadcasting with radio waves is happy to be overheard; if you are using a laser there's a chance you are trying to keep your communications secure. As ever with SETI, the slim chance of success is outweighed by the extraordinary advantages that detection would bring. What if we suddenly find ourselves plugged into the galactic internet? Imagine how many videos of Richard Dawkins arguing with creationists we could watch then.

As well as *Breakthrough Listen*, Milner also teased a second, equally enticing prospect called *Breakthrough Message*. Although full details have yet to be announced, the gist is that a million dollars in prize money awaits anyone who can create a digital message that will 'represent humanity and planet Earth'. As Ann Druyan explained, 'the *Breakthrough Message* competition is designed to spark the imaginations of millions, and to generate conversation about who we really are in the universe and what it is that we wish to share about the nature of being alive on Earth'.

And if all that's not enough to get your head spinning, try this: all of the data will be available online, as will any code that SETI writes to crunch it. If you want to get in on the act by analysing the data yourself, or by pimping *Breakthrough Listen*'s software, you are most welcome. Or if that seems a bit hands on, you can just download the SETI@home app and allow it to piggyback your laptop's spare processing power, becoming yet another node in the world's largest supercomputer. As a man who made his money via social networks like Facebook, Milner is clearly intent on making *Breakthrough Listen* a club that everyone wants to join.

ANOTHER EARTH

The timing of Milner's announcement couldn't have been better, coming just days after the Kepler Space Telescope found Kepler 452b, the planet they are calling 'Earth 2.0'. It would probably be more accurate to call it 'super-Earth 2.0', because it's actually 60 per cent bigger in diameter than our home planet, and would have twice the surface gravity. If it has water, it would have all the exciting life-friendly features of super-Earths that we mentioned in the last chapter, with Indonesian-type volcanic archipelagoes surrounded by shallow seas.

In that case, of course, we were talking about super-Earths orbiting close to orange stars; Kepler 452b, on the other hand, is orbiting in the habitable zone of a yellow star like our own Sun.[3] Like all of the Kepler planets, it's quite some way away; 1,400 light years away, in fact.[4] In other words, if we did intercept a message, they would have to have sent it 1,400 years ago, just as if they were to receive one from us tomorrow we would have had to have sent it in AD 615, fifteen years before the Prophet Muhammad's conquest of Mecca.

On the other hand, the star that Kepler 452b is orbiting is six billion years old. Again, let's assume that the Earth is typical, and, on average, intelligent communicable life arises on Earth-like planets after four billion years. Remembering that, thanks to gamma-ray bursts, no life would have been possible on any planet before five billion years ago, we can conclude that life on Kepler 452b could have a billion-year head start. What will life on Earth look like after another billion years of evolution?

I'm not one for predicting the future, so let's do our usual trick of falling back on what we already know. One billion years of evolution on Earth takes us back before the Cambrian explosion, to a time when life was, by and large, single-celled. No trilobites, no sponge-like Ediacarans, just the odd brightly coloured microbial mat. It's an unsettling thought, but whatever we find on Earth in another billion years might easily be as different from us as we are from bacteria. Martin Rees proposes that such creatures won't even be carbon-based life forms as we know and love them, but the robots of a long-dead civilisation.

---

3 Kepler 452b makes an orbit every 385 days.

4 Kepler is pointed at a distant but dense portion of the starfield between the constellations of Lyra and Cygnus, where it monitors the brightness of around 150,000 stars, all of which are between several hundred and several thousand light years away.

Whatever the case, it's becoming clear that, when it comes to communicating with aliens, there's a timing problem. They may be out there, but are their signals reaching us right now, and are we capable of understanding them? When Frank Drake formulated his famous equation to calculate the number of communicable civilisations, a crucial factor was the length of time for which a civilisation is detectable. But there's a world of difference between detectable and decipherable. We might be able to receive radio broadcasts from Kepler 452b – albeit via some extraordinarily poky transmitters – but how would we ever be able to decode them?

Let's imagine for a moment that *Breakthrough Listen* is wildly successful, and in seven years' time, after searching 752,656 stars, we finally detect a laser signal from an Earth-sized planet orbiting a Sun-like star 355 light years away.[5] Just like in the movie *Contact*, there's a recognisable call signal – prime numbers, perhaps, or the digits of π – followed by a short broadcast. How will we know if that broadcast contains a message, rather than just being random noise? And if it does contain a message, how will we translate it?

Strangely enough, this isn't the first time that scholars have confronted such a question. For centuries, some of the greatest minds in Europe struggled to understand the sacred texts of a distant people, convinced that they might contain wisdom that would speed the technological and spiritual progress of mankind. So many had tried and failed that whoever managed to crack the code was guaranteed immortal glory. Against the odds, one young linguist

5 The website for the Institute for Computational Cosmology at Durham University gives 250,000 stars within 250 light years, so I've assumed even density of stars to get roughly 750,000 stars within 360 light years.

succeeded, becoming a French national hero. We are about to meet the brilliant Jean-François Champollion.

THE PAST IS AN ALIEN COUNTRY

At the time of the French Revolution, if you were seeking answers to the big questions in life beside the Bible there was really only one place to look; the secrets of the Ancient Egyptians. Not only was the Hebrew story entwined with that of the pharaohs, but there were hints that Egyptian civilisation had equally divine roots. There was a catch, however. While a few Ancient Egyptian artefacts had found their way to Europe during the time of the Roman Empire, Egypt itself had long been out of bounds.[6] And even more tantalisingly, the script on those artefacts, known as hieroglyphs – from the Greek for 'sacred writing' – had been indecipherable for the best part of fourteen centuries.

Thanks to Napoleon Bonaparte, that was about to change. In an attempt to emulate his hero Alexander the Great, he decided to colonise Egypt, with the added intention of digging a canal through Suez that might renew France's interest in India. On 1 July 1798, after evading Nelson's fleet in the Mediterranean, he landed near Alexandria with over 400 ships and 38,000 men.[7] Among them were the cream of French intellectual society, such as the mathematician Joseph Fourier and the naturalist Etienne Geoffroy

6 Since the Islamic conquest of Egypt in AD 641, in fact.

7 Precise figures are hard to come by, but, roughly speaking: 30,000 infantry, 3,000 cavalry and 3,000 artillery and engineers, plus Napoleon's personal bodyguard of 380. Of the rest, 300 were women and 167 were savants. As well as 400 transport ships, Napoleon had thirteen ships of the line and seven frigates. Nelson had thirteen ships of the line but – following a storm in the Med – no frigates.

Saint-Hilaire,[8] spread among seventeen different ships for safe-keeping. While Napoleon ultimately failed to subjugate Egypt, it was these so-called savants who were to bring back the real prize: the Ancient Egyptian artefact known as the Rosetta Stone.

THE VALLEY OF THE KINGS

After Alexander the Great had conquered Egypt in 331 BC, the country had been ruled by the Ptolemaic Dynasty, a rather decadent line of Macedonian noblemen who had eventually assumed the role of pharaohs. It was Ptolemys I to III who built the world's first library at Alexandria, for example, and Ptolemy XIV who married his sister, the famous Cleopatra VII, only to be cuckolded by Julius Caesar.[9] Following the Islamic conquest, Egypt was ruled by a succession of Islamic Caliphates and Sultanates, the last of which was that of the Mamelukes, which lasted from 1250 to 1517.

At that point the Ottomans took over, controlling the country from Constantinople but retaining the Mamelukes as an aristocratic ruling class. By the time of Napoleon, however, the Mameluke leaders Ibrahim Bey and Murad Bey had become increasingly powerful, disrupting trade and generally doing little to ingratiate themselves with the Ottoman Sultanate. It was this power vacuum that Napoleon intended to exploit, presumably relying on the fact that France's erstwhile allies the Ottomans were cheesed off enough with the Mamelukes to remain neutral.

8 Whom we last met courtesy of our mutual friend the duck-billed platypus.

9 I'm not saying that building a library is decadent; it's more the incest that I'm aiming for. That said, the practice of brothers marrying sisters – and worse – was common in the upper echelons of Ancient Egyptian society.

To begin with, he was largely successful, easily capturing Alexandria and Rosetta on the coast, then defeating the Mameluke forces of Murad Bey at Shubra Khit on the Nile on 13 July 1798, and taking Cairo following the Battle of the Pyramids on 21 July. It wasn't to last. Barely two weeks later, on 1 August, Nelson finally caught up with the French fleet at Aboukir Bay (otherwise known as the Battle of the Nile) and delivered one of the most crushing naval victories in recorded history. No British ships were lost, while eleven of Napoleon's thirteen ships of the line were sunk. Napoleon had his colony, but they were a long way from home.

WHAT'S IN A NAME?

While Napoleon battled both the British and the Mamelukes, his savants were conducting just as earnest a campaign on the cultural front. Their first Ancient Egyptian find, at Alexandria, were two obelisks, one standing and one fallen, both of which were covered in hieroglyphs. Nicknamed 'Cleopatra's Needles' by the savants,[10] each contained a high proportion of what Napoleon's soldiers called *cartouches*, or 'gun cartridges'; namely, groups of hieroglyphs encircled with an oval and abutted by a line. For example, towards the top of Cleopatra's Needle in London you can clearly make out the following:

10 Or, rather, les Aiguilles de Cléopâtre. The supine one now resides on the Thames Embankment in London, the other in New York's Central Park. Both were gifts from the Egyptian government decades after Napoleon's departure from Egypt. The connection with Cleopatra is that she designed the Caesarium, or 'Palace of the Caesars', at Alexandria, and after her death the Emperor Augustus transported the needles from outside the Temple of the Sun in the city of Heliopolis, near modern Cairo. They were originally commissioned in 1450 BC by Pharaoh Thutmose III to honour the sun god Ra.

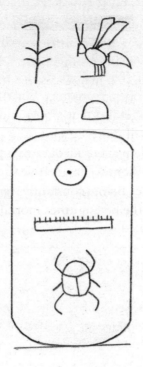

To put you out of your misery, this is the so-called throne name of the 18th Dynasty Pharaoh in whose honour the obelisks were commissioned, the great military hero Thutmose III, who ruled Egypt from 1479 to 1425 BC.[11] In appropriating the needles for Caesar's monument, Augustus certainly knew what message he was sending out to the Egyptians.

Thanks to Augustus and others, who brought many obelisks back to Rome as trophies, cartouches had been known to Western scholars for centuries. The belief at the time was

---

11 The 18th Dynasty, by the way, is the one that belongs to another famous pharaoh, Tutankhamun, who reigned from (roughly) 1332 to 1323 BC.

that they were made up of what we call ideograms, that is to say, pictures with a symbolic meaning. The sedge ideogram, for example, that sits above the cartouche for Thutmose III, represents Upper Egypt. The bee, on the other hand, signifies Lower Egypt. The semicircle, for reasons best known to the Egyptian scribes, represents 'Lord' or 'King'. Putting it all together, you get 'Lord of Upper and Lower Egypt'; or maybe 'He who will unite Egypt'.

Part of the problem about ideograms is that you can argue about them all day. Some scholars have suggested, for example, that the sedge, being constantly renewed, represents eternal life, while the bee signifies mortality. It may be that as far as the Ancient Egyptians were concerned, those four hieroglyphs contain all of those meanings and more. The cartouches of Alexandria, however, were nothing compared to the riches that awaited the savants in Upper Egypt.

THE FOUNDATION STONE

If Napoleon was disheartened by the defeat of the French fleet at the Battle of the Nile, he wasn't letting on. Apparently undaunted, on 22 August 1798 he inaugurated the Egyptian Institute of Arts and Sciences, and installed its member savants in one of the Mameluke palaces of Cairo. The work of the Institute was to be published a few years later as the *Description de l'Egypte*, and launched a wave of Egyptomania in Napoleonic France every bit as feverish as the one that had once gripped Ancient Rome.

While Napoleon's generals attempted in vain to engage Murad Bey in the deserts of Upper Egypt, one particular savant accompanied them in the hope of discovering further Ancient Egyptian monuments. Dominique Vivant,

Baron Denon was an artist who had been big news at the court of Louis XV, but had managed to avoid proscription, eventually befriending Napoleon via Josephine's Paris salon. At Dendera he encountered the extraordinary Temple of Hathor, with its every wall and ceiling covered in Ancient Egyptian inscriptions. They contained far fewer cartouches, and a profusion of new, as yet unknown hieroglyphs.

Returning to Cairo in the middle of August 1799, Denon impressed upon his fellow savants the vital urgency of translating the hieroglyphs. Any lingering doubt as to whether these elegant drawings were literal or symbolic was gone; all the wisdom of Ancient Egypt was suddenly theirs for the taking. More savants – Fourier included – were dispatched to Dendera to make further copies, while others attempted to decipher those in Denon's drawings. Yet with no way of knowing the subject matter, there was no way to begin a translation. Just days later, all that changed with the surprise arrival in Cairo of a large lump of dark grey granite.

TONGUE AND GROOVE

As luck would have it, a mere month earlier, on 19 July 1799, a group of French soldiers had been hard at work strengthening the defences of Fort Rashid, an Ottoman-built outpost at the mouth of the Nile near Rosetta. As the British now controlled the Mediterranean, this was to be a vital link in the defence of the Nile, and hence the entire colony. An ancient wall was being demolished when a soldier called D'Hautpoul spotted a grey slab with some sort of inscription on the side. The stone was passed up the chain of command to an officer named Michel Ange Lancret, who had been recently elected to join the Institute of Egypt.

On examining the stone, Lancret saw that the inscription was made up of three different scripts. One was Ancient Greek, one Egyptian hieroglyph, and the third he failed to recognise. Working on the Greek script, he was able to see that the text was a fairly workaday decree by the priests of Memphis detailing the good deeds of Ptolemy V and how precisely he was to be honoured. Dull though it might appear, Lancret immediately grasped its immense importance. If the text of all three scripts was the same, the savants finally had the means of decoding Ancient Egyptian.

IDIOT SAVANTS

Sadly, no sooner did the savants have the stone than it slipped through their grasp. Following the French defeat at the Battle of the Nile, the Ottomans got off the fence and sided with the British against Napoleon. With his hold on the colony looking increasingly shaky, Napoleon first repelled Ottoman forces in Syria, and then marched to the coast, where on 25 July 1799 he led a decisive defeat of the Ottomans at the Battle of Aboukir.[12]

While exchanging prisoners following the battle, Napoleon learned from the British that the political situation in France had deteriorated, and that the Directoire Exécutif were facing a potential *coup d'état*. Sensing his time was at hand, Napoleon hastened to Alexandria and set sail for France on 22 August, leaving the colonists in the lurch. Back in France, and buoyed by his recent victory over the

12 The very bay, in other words, where the French had lost the Battle of the Nile in 1798.

Ottomans, Napoleon was able to stage a coup of his own, and by 9 November he was installed as the First Consul; by 1804 he was Emperor.

Needless to say, Egypt soon fell to the Ottomans and the British, and the savants were placed in the undignified position of having to barter their way back to France. One of the many treasures they were forced to surrender was the Rosetta Stone, which was acquired by the British and transported to the British Museum in London where it has been on display more or less ever since. The savants, of course, had their copies of the Rosetta Stone, made by applying ink directly to its surface and using it as a printing block, and which they eventually presented over three plates in the *Description de l'Egypte*.

CLEOPATRA'S NOODLES

This, in case you were wondering, is the point where Jean-François Champollion comes in. A few months after his return from Egypt, the great mathematician Fourier took up residence in Grenoble. Like the other savants, Fourier styled himself as '*Un Egyptien*' and remained obsessed with all things ancient, including the hieroglyphs. When inspecting one of the local schools, he was so impressed by the linguistic abilities of a twelve-year-old student named Champollion that he invited him to his study to show him some of his Egyptian antiquities. Learning that none of the inscriptions could be understood, Champollion declared his intention to decipher the hieroglyphs.

Not that it was easy. Part of the problem, as Champollion was to discover, was that the third script on the stone, known as Demotic, was virtually unknown in academic circles. With little progress being made with the hieroglyphs themselves, the logical step seemed to be the translation of the Demotic, but that too proved a stubborn problem. A later form of Egyptian script, Coptic, was better known, and appeared to be derived from Demotic, and Champollion made it his business to become fluent in it.

Unknown to Champollion, however, he had a rival. Thomas Young was a trained physician and polymath, probably best known for proving that light is a wave by means of an experiment known as Young's Slits.[13] Fluent in a baffling number of languages, Young was late to the Rosetta Stone but quickly made up for lost ground. Focusing on the cartouches as the means to crack the code, he made a thrilling deduction. Knowing that the Ancient Greek and the Demotic scripts contained the name of Ptolemy, he set about trying to find the relevant cartouche among the hieroglyphs.

His reasoning was simple. Although believing, as everyone did at the time, that hieroglyphs were ideogrammatic, while the Demotic script was phonetic, he couldn't help noticing that some of the Demotic letters appeared to be derived from hieroglyphs. Could it be that some of the hieroglyphs represented sounds rather than ideas? If that was true, then those sounds would almost certainly be used to spell out a foreign name like Ptolemy. Sure enough, he found a cartouche which seemed to do the trick, and assigned the following letter sounds:

---

13 Young showed that if a light source was obscured but for two parallel slits, the light from those slits would produce an interference pattern. This landmark experiment was subsequently modified by one of my all-time physics heroes, G.I. Taylor, to prove the quantum nature of light. Basically he turned the brightness of the lamp down until there could only be one photon of light in the system at any one time, and went on a sailing holiday. When he returned, there was an interference pattern, just as in Young's experiment, proving there was no such thing as a 'single path' for a quantum object. Small things, in other words, can be in two places at once; in fact they can be in all places at once.

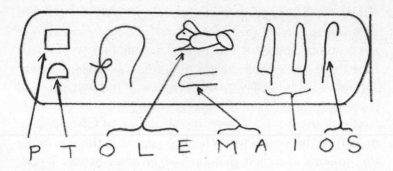

Unfortunately for Young, he remained convinced that the Egyptian names — and the rest of the hieroglyphs — were ideograms, not phonograms, and so made little further progress. Reading of Young's success, however, Champollion was inspired. Counting the characters on the Rosetta Stone, he found that 486 words of the Greek script were matched by 1,419 hieroglyphs. Grouping the hieroglyphs as best he could, he found that the total number of these 'words' was roughly 180. Clearly something was amiss. Could it be that Ancient Egyptian was much more complex than anyone had imagined, and contained a mix of 'picture' glyphs and 'sound' glyphs?

The final key was provided not by the Rosetta Stone, but by an obelisk obtained by the British adventurer William Bankes. Discovered in the Temple of Isis on the sacred island of Philae in the Nile, near Aswan, the fallen obelisk and the base from which it had become detached had already made an appearance in the *Description de l'Egypte*. When it arrived in England in the summer of 1821, Bankes realised that a Greek inscription on the base contained the names of Ptolemy VIII and Cleopatra III, and that one of the two cartouches within the hieroglyphs matched that for Ptolemy in the Rosetta Stone. Could the other cartouche be that of Cleopatra?

Excitedly, Bankes had a lithograph printed of both the Greek and Ancient Egyptian inscriptions, and sent a copy to Young. Unable to make any progress, Young decided that the copy was inexact and abandoned any attempt at translation. Another found its way into the hands of Champollion in France, however, who checked first that the cartouche for Ptolemy matched that on the Rosetta Stone. It did. Disregarding Young's original letter sounds, Champollion came up with the following:

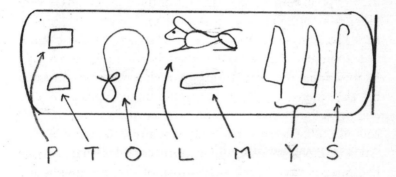

Turning to the second cartouche, he immediately recognised four of the glyphs:

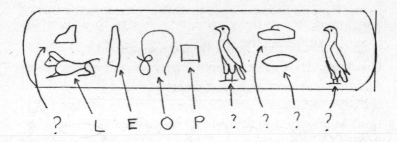

Assuming that the others must also be phonograms, and that the same sound could be represented by more than one phonogram, he assigned them the following values:[14]

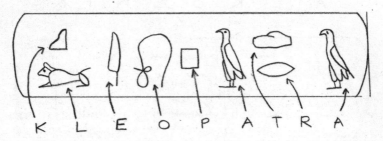

He then applied himself to a third cartouche:

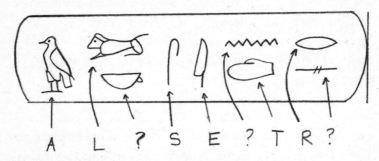

14 There's an added bit of brilliance here; hieroglyphs don't usually contain vowel sounds, and neither does Demotic script. Vowel sounds came in with the Egyptian–Greek combo that is Coptic.

Knowing six of the glyphs, he was able to deduce the name of Alexander the Great:

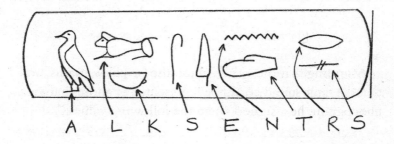

By 1824, Champollion had cracked virtually the entirety of hieroglyphics. Publishing his findings in the sensational *Précis du Système Hiéroglyphique*, he showed that there were actually three types of glyphs: ideograms, phonograms and determinatives. It was phonograms, not ideograms, that made up the heart and soul of the Ancient Egyptian hieroglyphs; in fact, there was some interplay between the two, since the consonants that made up an ideogram could also be used as a form of phonogram.

To give one example: the ideogram for a scarab beetle, as in the case of Thutmose III's cartouche, can represent the idea 'that which will be' or the three consonants 'hpr'. This was confusing even to the Ancient Egyptians, which is where determinatives come in. Placed after a phonogram, they let you know the sense of a word. The phonograms for chmplln, for example, might be followed by the determinative for a man 𓀀, showing that they spelled 'champollion'.[15]

WE ARE NOT ALONE

Contrary to the hopes of Champollion and others, our studies of the Ancient Egyptian texts have not made us adept in divine secrets, or shown us how life originated on Earth. Instead, they have intimately connected us with one of the first civilisations, revealing both its cruelty and its wisdom, its economic power and moral weakness. Considering that we are separated by over five thousand years of cultural evolution, we and the Ancient Egyptians are remarkably alike. The closeness of that bond is not only rewarding in its own right, but hints at another, equally close kinship with the humans that migrated to the Levant from sub-Saharan Africa. But for writing, we are all strangers.

The worldwide mania that followed the discovery of Ancient Egyptian hieroglyphics no doubt has its parallels in how the international community would react to the detection of an alien signal. There will be people who expect it to contain the secrets of the universe, of technologies beyond our wildest imagination: the final fulfilment of our blocked wish for unbounded energy, perhaps; time travel; maybe even an end to world poverty. What we find in it will almost certainly be something different. Whatever it

---

15 So what do the sun ☉, the gaming board 𓏤𓏤𓏤𓏤 and the scarab 𓆣 mean in the throne name of Thutmose III? The sun ☉ meaning 'that which is' is easy enough, and we already know that the scarab denotes 'that which will be', but what about the gaming board? Crucially, it's not an ideogram, but a phonogram, 'mns'. 'Menes' was the first pharaoh of Egypt, flipping us back to an ideogram with the meaning 'that which was'. Altogether we get 'present, past, future' or 'everlasting'. So Thutmose's throne name is 'Son of Ra, He who is everlasting'. Phew.

is, like the deciphering of the hieroglyphs, it will dissolve many of the physical boundaries between our cultures. In short, we will not be alone.

Like the hieroglyphs, however, any message we do detect will be unbearably hard to decode. Ancient Egyptian society, being more technologically primitive than ours, might have produced a more simple form of writing, but it did not. The complexity of the hieroglyphs speaks to a different kind of literacy to that we are used to today; one where the very act of writing was a divine act, where symbolism and literalism collided as violently as they do in, say, *Ulysses* by James Joyce. Modern English, for example, readily submits to digitisation, becoming a string of 1s and 0s with little lost in the way of meaning. Could such a highly visual language as Ancient Egyptian be encoded as faithfully? Our languages have become more pliant, certainly, but drained as they are of visual symbolism, have they become less poetic?

Finally, the great lesson of the hieroglyphs is the crucial part played by the Rosetta Stone. Without knowing the content of at least some part of a message, we have no way to decode it. When we talk to aliens, what will we use in its place? After all, in the case of the hieroglyphs we shared enough culture with the Ancient Egyptians to know what a scarab beetle looked like, for example, or that the Sun takes the shape of a circle. What do we do when the life forms on the end of the phone are giant centipedes that communicate via chemical odours, or gelatinous-tendrilled air-dwelling balloons that signal using bioluminescence?

RADIO GA GA

Thankfully, there is one obvious candidate for a Rosetta Stone, and we don't need to go digging for it in the dunes of a future desert planet: the physics and mathematics required to build a radio telescope. We may not share body parts, or culture, or even the same biochemistry, but in order to send us a radio message our alien callers will at least share our penchant for radio technology. And that in turn means they will have at least the same understanding of mathematics and physics as we do.[16]

Just as we assume the aliens will broadcast in the least noisy part of the spectrum, we are fairly safe in assuming that they will begin their message with some form of Rosetta Stone. The most common element in the universe is hydrogen, so why not start there? They could kick off with the charge on the electron, for example, closely followed by the mass of a proton, then seal the deal with the speed of light. The problem, of course, is that to put a number to such things, intelligent creatures need a system of measurement. That in turn requires some kind of unit, and it's not that likely that the Kepler 452f-ians are using the metre, second and kilogram. Which is where some rather ingenious things called dimensionless numbers come in.

A dimensionless number is one that has no units, and is therefore the same whatever measurement system you choose. You already know one: pi, being the circumference of a circle divided by its diameter. Pi crops up everywhere

16 Granted, we are making the assumption that mathematics is universal, and not just something we cooked up to try and make sense of life on Earth. So in a sense we are funnelling the number of communicable civilisations down to 'those that have radio dishes and share our mathematics'.

from General Relativity to Quantum Mechanics, and would no doubt be as handy to alien mathematicians as it is to earthbound ones.[17] Most people could probably tell you the value of pi to three decimal places, being 3.142, or that it is approximately equal to $^{22}/_7$. But to physicists there is another number that is just as famous: 137.

That – or, rather, its inverse, $^1/_{137}$ – is the approximate value of what is known as the fine-structure constant, α. This enigmatic number describes the tendency for an electron to emit or absorb a photon, and while it is fundamental to a quantum mechanical description of electromagnetism – and therefore, the manufacture and operation of radio dishes – no one has the faintest idea where it comes from. And it's not alone. The gravitational coupling constant, α_G, would also be as familiar to alien physicists as the name of Cleopatra to the Ancient Egyptians. To those we can add such gems as the ratio of a proton mass to an electron mass, and the ratio of a neutron mass to a proton mass, all of which will be well known in alien worlds just as they are here.

Even with a Rosetta Stone in the form of basic physics, decoding the rest of an alien radio message will be fiendishly difficult. After all, it took some of the brightest minds on the planet twenty-three years to effect a translation of the hieroglyphs. To be successful we will need both brilliant scientists like Thomas Young, and equally gifted linguists such as Jean-François Champollion. The incipient age of interstellar communication will require a coming together not just of the sciences, but also the arts.

In the meantime, however, what do we look for? The practical answer is simple: any kind of transmission for

---

17 Pi, like many dimensionless numbers in physics, is irrational, meaning it can't be neatly expressed as a fraction.

which there is no known natural source. Like Jocelyn Bell Burnell, we will turn up a few pulsars along the way, but we will be all the richer for that. For the savants, the Ancient Egyptian monuments of Rome hinted that somewhere out there lurked the collected wisdom of an entire civilisation. In the quest for extraterrestrial intelligence, we ourselves are that hint. Assuming we do find some kind of unidentified transmitted signal, what then? Amazingly, even with no Rosetta Stone, all is not lost. And the story starts in the most unlikely of places; with *Ulysses* by James Joyce.

THE BLOOMSDAY BOOK

First serialised in the American journal *The Little Review*, James Joyce's *Ulysses* is rightly considered the masterpiece of literary modernism.[18] Everything about it defies convention. Its narrative, if you can call it that, might be summed up as 'two men go for a walk in Dublin, and very little happens'. Virtuosic, untrammelled, ruthlessly academic and intentionally obscene, *Ulysses* is about as idiosyncratic a text as you can imagine, dispensing with all the pre-existing conventions of character, speech, style, comprehensibility and believability. It is exhilarating to read, at the same time unlocking the gilded cage of form and imprisoning you with its vaulting genius.

Yet as anarchic and free-flowing as it appears, when it comes to its underlying structure, *Ulysses* is subject to the

18 *Ulysses* was serialised from March 1918 to December 1920, then published in book form in Paris in 1922. Its title is the Latin name for Odysseus, the hero of Homer's epic poem the *Odyssey*; like his editor Ezra Pound, Joyce wasn't afraid of an allusion or two.

most rigid of rules. As first noticed by the Harvard linguist George Kingsley Zipf in his 1949 masterpiece *Human Behavior and the Principle of Least Effort*, when submitted to statistical analysis, *Ulysses* is indistinguishable not only from one of the texts to which it alludes,[19] Homer's *Iliad*, but also to the Old English poem *Beowulf*, four Latin plays of Plautus and the language of the Plains Cree Indians. To cut to the chase, if you take all the different words in *Ulysses*, count how often they occur and then rank them in order, you start to see something extraordinary.

To see what I mean, take a look at the chunk of *Ulysses* that appears at the beginning of this chapter. To really see the pattern, you need to analyse the text as a whole, but all we are after here is the gist. It's fairly easy to see which word crops up the most: it's the rather unprepossessing 'of', which I count as appearing twenty-one times. Next is the equally un-Joycey 'the' which crops up eleven times. After that we have 'in' at seven times, and 'and' at five times . . . by which point you may have picked up the pattern. The second most frequent word occurs half as many times as the first most frequent; the third most frequent, a third as many; the fourth, a quarter. In short, if we define the *rank* of a word to be its place in the pecking order, and its *frequency* to be the number of times it appears, we can write:

rank = constant/frequency

It surprises me all over again just telling you, so to convince us both let's take a look at the data that Zipf presents in *Human Behavior*. This is the point at which a table comes in handy, so here goes:

19 The other, of course, is the *Odyssey*.

Word	Rank	Frequency
I	10	2,653
say	100	265
bag	1,000	26
orangefiery	10,000	2

Even more strangely, it doesn't end with written texts and spoken languages. Turning to data from the Sixteenth United States Census, conducted in 1940, Zipf showed that the same relationship applied for the population of cities, the number of shops within a city and the wages of the citizens within them. In other words, the tenth most populated city had a tenth of the population of the first most populated, and the one hundredth richest man had a hundredth of the wealth of the richest. What could all of this possibly mean?

To this day, no one is exactly sure. Anyone and everyone in practically every field of the humanities has had a crack at Zipf's law, and no one explanation has yet garnered a decisive following.[20] As the title of his book suggests, Zipf's own suggestion was that it is something to do with that universal human maxim 'anything for an easy life'. Why go out of your way to shop on the high street when the supermarket on the outside of town has everything that you need and more? Why hire another actor when Tom Cruise is available? Why use DuckDuckGo when you've got Google?

---

20 OK, this is the most nerdy joke in the entire book but here goes: when we rank the various explanations of Zipf's law together with the number of academics that subscribe to that explanation, we get the only data set involving human decision-making which isn't subject to Zipf's law. Snarckle.

Of course, when it comes to language, two things quickly make you lose the will to live. The first is when too many sounds are repeated, as anyone who has been on a long car journey with a three-year-old can attest. The other, as you will know from trying to speak French on holiday, is when there are too many different sounds to be able to tell them apart. Zipf saw language as a trade-off between these two extremes, a middle ground where the balance – weighted somewhere towards shorter words that are easy to say and understand – was just about perfect. Whatever the case, it seems that Zipf's law is a necessary-but-not-sufficient property of language. And most fascinatingly for our story, in 1999 three researchers named Laurance Doyle, Brenda McCowan and Sean Hanser found that Zipf's law applies to the whistles of bottlenose dolphins.[21]

THE ORDER OF THE DOLPHIN

SETI has a long history of research into dolphin communication. One of the ten attendees at the first SETI conference at Green Bank in 1961 was the neuroscientist John Lilly. His book of that year, *Man and Dolphin*, had been an international bestseller, and claimed not only that dolphins were capable of complex emotions, but that they might also be capable of speaking human languages.

Lilly's book had caught the attention of Frank Drake, who wanted to understand the potential challenges of communicating with other intelligent species. Lilly, with his boundless charisma and movie star looks, was an instant

21 You can find a fascinating discussion of all things Zipf in Alex Bellos' rather brilliant book *Alex Through the Looking Glass*. Thanks Alex for bumping into me in a coffee shop and handing me a copy at the very moment I happened to be writing this chapter.

hit. Not only were the brains of bottlenose dolphins larger than ours, he informed the conference, but they were just as densely packed with neurons; in fact, parts of their brains looked even more complex than their human counterparts.

What's more, dolphins appeared to have their own language. Using tapes he had recorded at his new purpose-built Communication Research Institute on the island of St Thomas in the Virgin Islands, he demonstrated how bottlenose dolphins were able to communicate with one another with whistling sounds that they made using their blowholes. If he slowed the tapes down, he showed, the dolphins' squeaks and clicks even sounded like human language. Might it be possible to teach them to speak English?

Later, Frank Drake was to come to the reluctant conclusion that Lilly's work was 'poor science', and that he had probably distilled hours of recordings to find those little bits that made their speech sound humanlike. At the time, however, Lilly's findings were enthralling, providing just a taste of the non-human intelligence that they were all seeking. Only Philip Morrison offered a note of scepticism, observing that dolphins, intelligent as they were, couldn't build telescopes with flippers.

On disbanding, the group decided to call themselves 'The Order of the Dolphin'. A few weeks later, Frank Drake received a small package in the mail from Melvin Calvin.[22] An identical package was addressed to Struve, and Drake later learned that each conference participant had received one. Inside the box was a silver badge, a replica of an Ancient Greek coin in the shape of a leaping dolphin. The badge was a reminder that not only had they formed their own fraternity, but that in admiring

22 For his work on the Calvin Cycle, the process by which plants photosynthesise sugars using the light energy captured by chlorophyll.

the intelligence of dolphins they were honouring an academic tradition that went back to the Ancient Greeks.

THE GIRL WHO TALKED TO DOLPHINS[23]

The interest from Frank Drake, Carl Sagan and others helped Lilly to secure NASA funding for his Communication Research Institute, and in 1965 he conducted one of the all-time strangest scientific experiments, when for ten weeks he agreed that a twenty-two-year-old researcher, Margaret Howe, should be isolated with a young male bottlenose dolphin called Peter. Howe had been attempting to teach Peter to speak by repeating English letters, numbers and words and trying to get Peter to 'say' them back to her, much as a mother might teach a child to speak. Howe became convinced that she would make more progress if she was in constant contact with Peter, and persuaded Lilly to allow her and the dolphin to cohabit.

The upstairs of the house was plastered and flooded so that Howe and Peter could share the same living space, with the knee-deep water being just shallow enough for Howe to wade through, and just deep enough for Peter to swim in. Howe slept on a foam mattress in the middle of one of the pools, worked at a desk suspended from the ceiling and ate tinned food to minimise contact with outsiders. The days were run to a strict timetable, with Lilly giving precise instructions as to what Peter was to be taught and how it was to be documented. Howe was extremely dedicated, cropping her hair and even at one point painting the lower half of her

---

23 I highly recommend that you check out the film of the same title by Chris Riley, he of *In the Shadow of the Moon* fame, which contains original footage of the experiment and a present-day interview with Margaret Lovatt.

face in thick white make-up and applying black lipstick so that Peter might more clearly see the shapes her lips made.

Dolphins are promiscuous by nature, and as the experiment went on, Peter began to show a sexual interest in Howe; by week five she records in her notes that 'Peter begins having erections and has them frequently when I play with him'. When she rebuffed his advances, Peter would become aggressive, using his flippers and nose to bruise her shins. When his conjugal visits with the other dolphins at the facility became so frequent as to be disruptive, Howe decided to relieve him manually so that he might focus on his lessons. Lilly, whose judgement may have been clouded by his incipient interest in LSD, noted, 'I feel that we are in the midst of a new becoming; moving into a previous unknown . . .'

Gregory Bateson, the distinguished linguist who served as the Institute's director, and who was leading the research on dolphin to dolphin communication, was less than impressed. In his view, Peter was simply mimicking Margaret's speech in order to get fish, with little or no comprehension of what was being said. Carl Sagan, too, was sceptical that any progress was being made and suggested that Lilly switch to experiments which could verify whether dolphins could convey information to one another rather than trying to teach them English.

Eager for results, Lilly became increasingly desperate, and in the summer of 1966 he took the drastic step of injecting two of the dolphins with LSD to see whether that might improve their language skills. Thankfully, the drug appeared to have no effect; not even when Lilly took a pneumatic drill to the rock at the side of the pool to mess with the dolphins' super-sensitive hearing. This new twist was too much for Bateson, who left the facility. Lilly's NASA funding was withdrawn, the Institute closed down and SETI was forever blemished by association.

JUMPING THROUGH HOOPS

The lesson of the Lilly experiments – apart from 'don't give dolphins handjobs or LSD' – is a basic one. When it comes to close encounters with other species, it's all too easy to try to impose our own ideas of how they should think, feel and behave. To us, a dolphin's lack of facial expressions, for example, might seem to indicate a lack of empathy; to a dolphin, our reluctance to stand on our heads when we greet them might appear equally inscrutable. On one count, however, Lilly does appear to have been correct; dolphins have all the trappings of high intelligence.

For example, we know that not only are bottlenose dolphins capable of understanding signs, but the order in which they are shown those signs – in other words, the syntax – makes a difference. In the 1980s, Louis Herman and his team at the Dolphin Institute in Hawaii successfully taught captive dolphins 'words' in the form of arm signals. Bottlenose dolphins were shown to be capable of understanding 'sentences' of up to five 'words', easily distinguishing between 'take the ball to the hoop' and 'take the hoop to the ball'.

As demonstrated by the work of Diana Reiss, bottlenose dolphins are capable of complex manipulation, creating underwater bubble-rings with their blowholes that they can then swim through; in one case a captive dolphin formed a stream of bubbles with its blowhole that it then shaped into a ring using its tail.[24] They also show self-organised learning, interacting with a large poolside keyboard where

24 Diana Reiss has some extraordinary footage of this in her TEDxBrussels talk entitled 'The Dolphin in the Mirror'.

they could request objects by pressing the relevant key, and exhibit self-awareness when presented with an underwater mirror, orienting their bodies to examine temporary marks made by Reiss and her fellow researchers.

What was less clear, up until the late 1990s, was whether or not bottlenose dolphins' wide range of vocalisations constituted any form of language. In the broadest of strokes, they make two main kinds of sounds with their blowholes: whistles and clicks. The clicks, generally speaking, are their equivalent of sonar, helping them to locate their prey and generally find their way around when visibility is poor.[25] It's the whistles – many of which can be outside the range of human hearing – where all the linguistic action appeared to be, but no one had any concrete proof.

That was until the collaboration between SETI's Laurance Doyle and UC Davis's dolphin researchers Brenda McCowan and Sean Hanser. As an astronomer at SETI, Doyle had an interest in whether the mathematical techniques that were being used to pick out planets in the Kepler data could be used to decode alien messages. In particular, he was interested in what information theory, as devised by Claude Shannon, could tell you about an alien radio source. As you may remember from Chapter Five, Shannon showed that the maximum amount of information, H, contained in a message of i letters, each with probability p_i, was simply:

$$H = - \Sigma \, p_i \, \log_2 p_i$$

In other words – and this is the only part you need to grasp – we can work out how much information could

---

25 Dolphins also make another type of sound called burst-pulses, which are like a rapid series of clicks. Like whistles, there is evidence that these are also social calls.

possibly be in a message *just by knowing the letters the message is made up of, and how often they occur.*

If that sounds incredible, it should: to my mind, it's a leap on the level of Einstein's Theory of Relativity or Schrödinger's Wave Equation. Let's think back for a moment to the example we looked at in Chapter Five, where our apple-scrumping message is made up of two 'letters', 'torch on' and 'torch off'. Crucially, we calculated that the information contained in the message was 1 bit, even though we had no idea what that information was, i.e. whether you were coming scrumping or not.

The team's idea was simple. They had no SETI signal to look at, so why not look at the messages of dolphins using information theory to see how much information they were capable of carrying? And if they were capable of carrying a lot of information, was it anything like as much as human language?

NAME, RANK AND NUMBER

Doyle knew McCowan through the Planetary Society, a US non-profit which maintains an avid interest in SETI. One of the problems with understanding dolphins is their prodigious acoustic ability: water is much better at transmitting sound than air, and a dolphin can hear higher frequencies than a bat.[26] To make matters even more complicated, dolphin whistles can be very short, lasting as little as a few tenths of a second, and change rapidly in pitch. As part of her PhD, McCowan had developed software that could not

---

26 A little brown bat (*Myotis lucifugus*) can hear frequencies up to 115KHz; a common bottlenose dolphin (*Tursiops truncatus*) is good up to 150KHz. Humans hear up to 20KHz.

only accurately sample the individual whistles but could then categorise them by type. Now that there was a way of identifying the different 'letters' in a dolphin message, could they be analysed using information theory?

One of McCowan and Hanser's recent papers contained a table listing a number of different dolphin whistles and how often they had occurred, and to get things rolling the team decided to analyse them with a Zipf plot. For each of the forty-odd whistles in the table, they noted rank and frequency. Next they plotted them on a graph.

Zipf's Law, you'll remember, can be written as:

rank = constant/(frequency)

or

rank = constant x (frequency)$^{-1}$

For reasons that shall soon become clear, let's rewrite this as

rank = (frequency)$^{-1}$ x constant

This kind of relationship is known in the trade as a power law, meaning simply that one variable (rank) is related to another (frequency) by a power index, in this case -1.

That's no fun to plot on a graph, so a standard trick in physics is to take the logarithm of both sides of the equation:

Log (rank) = log { (frequency)$^{-1}$ x constant}

log (rank) = $-$ log (frequency) + log (constant)

Which is of that famous form, y = mx + c, the equation of a straight line. Plotting the logarithm of each whistle's rank against the logarithm of its frequency, the team were astonished. It came out as a dead straight line with a backwards sloping gradient of –1.00, like this:

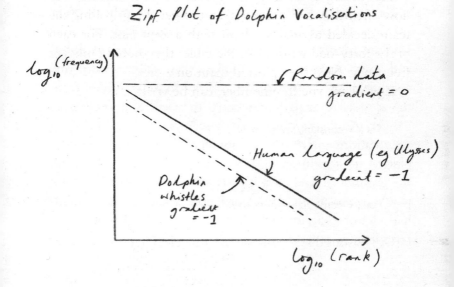

THE PLOT THICKENS

Unbelievably, dolphin whistles meet one of the basic requirements of symbolic language: like *Ulysses*, the *Iliad* and Plains Cree, they obey Zipf's Law. A result like that needs checking, so next the team looked at the Zipf plots of baby dolphins. Sure enough, the line ran with a flatter slope, indicating that a much wider variety of whistles was being used; in other words, the baby dolphins were babbling. Next, they plotted the whistles of dolphins between two and eight months old. This time, the Zipf line steepened to a −1.05 slope, showing that the toddler dolphins were repeating themselves. Finally, at between nine and twelve months, teenage dolphins plotted at −1.00, just like the adults. They had finally started making sense, no doubt earnestly informing one another that their parent dolphins were total losers.

Next, the team decided to check how the calls of other species compared with bottlenose dolphins. Squirrel monkeys are highly social New World monkeys from Central and South America which make alarm calls to warn one another of predators, and Belding's ground squirrels from the western United States chirp to warn one another of danger. Adults of both species were recorded, and their vocalisations analysed on a Zipf plot. The squirrel monkeys plotted at a less impressive but still noteworthy −0.75, and the ground squirrels at a measly −0.30. Whatever the Zipf plot was measuring, adult bottlenose dolphins and humans had a lot of it, squirrel monkeys had a bit and ground squirrels virtually none at all.

SO COME ON, DO DOLPHINS HAVE LANGUAGE?

As we've already mentioned, a Zipf plot is a necessary but not sufficient requirement for symbolic language, and it's fairly easy to see why. *Ulysses* follows a Zipf plot, but it would be a brave literary theorist indeed that claimed that much of the complexity in Joyce's masterpiece was captured by the knowledge of how often each word occurs.

Clearly there is a lot more going on, and luckily Shannon has a fair bit to say about that, too. To write *Ulysses*, you not only need the right words, but you also need to put them in the right order. Looking at the text at the start of this chapter, for example, we can see that 'of the' occurs five times; 'of our' occurs three times. These two-word chunks are known as 'digrams'. Clearly we need a lot more text to make a decent job of it, but you can see that we could do the same job for digrams as we did for individual words, working out the probability of each for *Ulysses* as a whole and plugging them into one of Claude Shannon's – admittedly fierce – equations.

If you do that, the number that you end up with is called a message's 'second order Shannon entropy'. In fact you can do the same sort of job with trigrams, quadrigrams, quintigrams, and so on, producing figures for the third, fourth, fifth, and higher Shannon entropies that capture more and more of the complexity of the message. Human language, for example, has a Shannon entropy of around eight, meaning that, on average, rules of syntax connect eight words at a time.[27] And the dolphin whistles? They plotted at four, round about the same number of symbols that Louis Herman showed dolphins were capable of handling when they learned sign language.[28, 29]

27 Another way of looking at this is: thanks to grammar, if you know seven words of a phrase you can have a go at guessing the eighth. Knowing eight words, however, won't help you with the ninth.

28 For the keeny-beany Shannonites: Zipf's law gives the distribution of word probabilities in the first-order Shannon entropy. There, I've said it.

29 I'm playing fast and loose with the words 'letter', 'symbol' and 'word' here, but I'm sure you get the point; Shannon's formulae apply whether you look at the text at the level of letters, phonemes, words, or phrases etc. That said, a higher Shannon entropy for letters will correspond to a lower one for words, e.g. first-order entropy for words of English corresponds to somewhere between fifth- and sixth-order entropy for English letters.

TALK TO THE ANIMALS

So that's it, right? Dolphins have language, but it's not nearly as complex as ours. The answer, as you might suspect, is 'not quite'. Firstly, we are measuring Shannon entropy, which tells us the *potential* for a message to contain information. Like a Zipf plot, a fourth-order Shannon entropy is a nec-essary-but-not-sufficient condition for complex language. Dolphins' whistles might obey complicated laws of syntax without actually meaning anything, although given their highly social behaviour that seems unlikely. Just as with the Ancient Egyptian hieroglyphs, until we manage to decipher bottlenose dolphin whistles we can't be sure.

And, second, a fourth-order Shannon entropy might not be the maximum entropy of dolphin whistles. As we know, to be able to calculate higher order entropies you need more and more text.[30] Doyle, McCowan and Hanser only had around 10,000 whistles to work with, so fourth-or-der entropy was the highest they could expect to find. You know you've reached maximum entropy when each *n-gram* – quadrigram, in this case – appears roughly the same number of times. Some dolphin quadrigrams, however, appeared a lot more often than others, which implies that if Doyle and his team had more whistle data they might have found higher orders of entropy. For all we know, dolphin whistles might have maxed out at eighth- or ninth-order, implying that – shock, horror – not only can they talk, but they are smarter than us.

30 As a rule of thumb, to calculate the nth-order Shannon entropy, you need 10n letters. In other words, for third-order entropy you need 1,000 letters; for fourth you'd need 10,000. *Ulysses* is approximately 265,000 words in length, so is capable of revealing anything up to fifth-order Shannon entropy, since $\log_{10}(265,000) \approx 5.4$

HOMO SAPIENS PHONE HOME

In my humble opinion, the work of Doyle, McCowan and Hanser has huge implications for any alien transmission we receive. Assuming we find a way to extract a message from it, one of the first things we are going to want to do is analyse it using information theory. That way, without knowing anything about what's in the message – 'We buy gold', for example – we can figure out its potential to carry meaning. A fifteenth-order Shannon entropy would have our cryptographers quaking in their Birkenstocks, but at least we would know what we were dealing with.

For us to be able to determine a fifteenth-order Shannon entropy, of course, we would need a whole lot of message. We'd also want it to be as diverse as possible, incorporating every conceivable kind of medium and exploring every nook and cranny of their culture. We'd want to hear alien music, flick through alien holiday snaps and watch alien movies. After all, without a Rosetta Stone decoding it is going to be prodigiously difficult. All we can hope for is some random cultural overlap – that we both spend long years raising children, for example, or share a love of selfies – so that we can find the equivalent of a cartouche: a small section of code for which the meaning is clear.

In short, we want the aliens to send us their internet.

METI-PHYSICS

So, finally, if we are to send a message, what should it be? In fact, should we even signal at all? Many eminent scientists, Stephen Hawking among them, believe we shouldn't. Contact with a more technological civilisation didn't turn out too well for the Plains Indians, he points out. To a certain extent, he's right; we don't know what's out there. Maybe an advanced civilisation will travel here by wormhole and suck up the Pacific into a giant water tanker.

As Seth Shostak, the director of SETI, recently pointed out in a *New York Times* article, that's a concern we never used to have. As we know, the Mir Message in 1962 set the ball rolling, with a brief Morse code message to Venus. Next came the so-called Golden Plaque message on *Pioneer*s 10 and 11, launched in 1972 and 1973. The following year we sent the most powerful message we have ever transmitted, the Arecibo Message, which was fired at a cluster of some 300,000 stars known as Messier 13. Depicting a stick man, a twist of DNA and a map of the solar system, the cruddy pixilation of the Arecibo Message makes the 1970s video game *Pong* look sophisticated.

That was followed by the *Voyager 1* and *2* probes in 1977 and the famous Golden Record. On it, as we've heard, Ann Druyan and Carl Sagan curated just about every kind of information they could get their hands on: speech, whale song, classical music, rock and roll, as well as images of the Taj Mahal and the underside of a crocodile. What they excluded, of course, was anything to do with war, politics or religion. After all, we didn't want to disappoint the aliens with our bad behaviour.

Since the Golden Record, the Crimea has become the main focus for Messaging to Extra-Terrestrial Intelligence, or METI as it is now known thanks to the Russian astronomer Aleksander Zaitsev. In 1999 and 2003 he supervised the sending of two messages known as the Cosmic Calls to nine nearby stars. And, not to be outdone, in 2008 NASA beamed the Beatles' song 'Across the Universe' at Polaris, also known as the North Star.

Rather than continuing to send such 'greetings cards' Shostak has a radical proposal, and one I heartily agree with: we should start sending 'Big Data'. As Laurance Doyle's work shows, we don't need to overthink this. The first thing the aliens will want to do is work out whether there might be any information in our message, and to do that they will need a lot of data. There are other good reasons, too. In sending them everything we've got, we are being as honest as we can be about who we are, warts and all. Let's not pretend we are sages, or saints. Let's be human.

We don't have to worry about overloading them. To a technological civilisation more advanced than our own, the world's several hundred exabytes of stored information is going to seem like a disk-on-key.[31] At the same time, while their technology might be more advanced, let's not assume that they themselves are necessarily smarter than us. It's our ability to manipulate information that has enabled our ascendance, not our individual smarts.

Let's send the internet. And while we're doing it, let's redouble our efforts to understand the communication systems of our fellow creatures. Until we can talk to dolphins we have little hope of being able to talk to ET. Just do one thing for me: somewhere up the front, let's stick James Joyce's *Ulysses*. Not only does it have maximum entropy to make your eyes water, but it's funny, and humour is something that has been sadly lacking in our messages so far.[32]

The supreme irony in all of this is: with the wacky hippy

---

31 Estimated in 2007, by Priscila Lopez of UOC and Martin Hilbert of USC, to be 2.9×10^{20} bytes.

32 Plus one of its best jokes is that it is impossible to understand.

inclusiveness of the Golden Record, Carl Sagan had it about right. He and Ann Druyan sent the internet of their day; now we have to do the same with ours. After all, as with all worthwhile communication the real message here isn't what we know about physics, or where we are in our solar system. It's the fact that we want to talk in the first place.

ARE WE ALONE?

We started this journey with a question. To answer it, first we needed to set a couple of things straight: the evidence for UFOs is weak, and the scientific credentials of SETI are strong. Next, we learned how the fundamental structure of the universe – the strengths of the interactions between its building blocks, for example – are fine-tuned to make carbon-based life a reality. All known life, we learned, is one; we are intimately related to every other organism on this planet, be it fungus, elk or bacterium.

To get a steadier grip on the slippery issue of detectable alien signals, we summoned the Drake Equation. Crucially we learned that there were a number of factors that combine to produce the overall probability that other Earth-like worlds might be calling. First, we needed to know the rate of formation of Earth-like planets. To calculate that, we needed to know the rate of formation of Sun-like stars, the fraction of those stars that have Earth-like planets, and how many Earth-like planets they have.

Thanks to the Kepler Space Telescope, these are questions to which we have some fairly accurate answers. Incredibly, those answers are remarkably close to those that the original Order of the Dolphin guessed at. Sun-like stars are formed at the rate of around one a year, and between a fifth and half of such stars have one Earth-like planet.

Next, the Drake Equation asked the rate of formation of life-bearing planets. Again, as per the original SETI meeting, our best guess is a large number; virtually the same as the rate of formation of Earth-like planets. Our evidence for this is the early emergence of microbial life, and the fact that it appears to have taken a vital first step – namely, the evolution of a cell wall – at least twice, giving rise to both the bacteria and their seabed-fellows, the archaea.

The next step was to get a feel for the likelihood of complex life, by looking at the major transitions that made it possible. Among them all, only one appears to be a bottleneck: the evolution of the mitochondria, which happened only once in our planet's 4.5-billion-year history, and without which the eukaryotic cell would never have got its start. Speaking personally, it's this stepping stone across the river of chaos that keeps me awake at night. Everything before and everything after seems to follow naturally, but without the vast reserves of energy provided by an archaeon enslaving a bacterium, it's hard to see how complexity might ever arise.

Returning to the Drake Equation, this 'eukaryotic bottleneck' provides our first departure from the early calculations of Drake and co. How can we put a number on the fraction of Earth-like planets that develop complex life? Granted, our investigations suggest that complex life inevitably develops intelligence, but with only one example of such life to go on, how can we conjure a number? We may not be in the dark, but we are definitely in the gloom. One in a hundred? One in a million, maybe?

To comfort us, we have the thought that there was at least one more instance of an endosymbiosis, that is to say when one type of cell enslaved another. That example, of course, was the co-option of the cyanobacterium by a eukaryote, creating the type of cell that we find in plants. Complex life may be rare, but like our own solar system, with its idiosyncratic Jupiter and absence of super-Earths, not that rare.

Finally, we learned that we, as humans, share civilisation, agriculture, language and culture with countless other beings here on Earth. If we are going to learn to talk to aliens, first we are going to have to learn to communicate with the other terrestrial and non-terrestrial intelligences on our own planet. Huge, ancient brains are out there, wanting to play, to commune and to teach. Are we alone? No. But it's down to us to make the first move.

Further Reading

All the books below are, in my humble opinion, not just great science writing, but great writing.

EXTREMOPHILES

The Voyage of the Beagle by Charles Darwin, Penguin Classics, 1989

Weird Life: The Search for Life That Is Very, Very Different from Our Own by David Toomey, W. W. Norton & Company, 2014

UFOS

Aliens: Why They Are Here by Bryan Appleyard, Scribner, 2005

The Demon Haunted World: Science as a Candle in the Dark by Carl Sagan, Ballantine Books Inc., 2000

SETI

The Eerie Silence: Searching for Ourselves in the Universe by Paul Davies, Penguin, 2011

Rare Earth : Why Complex Life is Uncommon in the Universe by Peter D. Ward and Donald Brownlee, Copernicus (December 10, 2003)

UNIVERSES

Just Six Numbers: The Deep Forces That Shape the Universe by Martin Rees, W&N, 2001

The Hidden Reality: Parallel Universes and the Deep Laws of the Cosmos by Brian Greene, Penguin, 2012

LIFE

What is Life? by Erwin Schrodinger, Cambridge University Press, 2012

Creation: The Origin of Life/The Future of Life by Adam Rutherford, Penguin, 2014

HUMANS

Human Universe by Brian Cox and Andrew Cohen, William Collins, 2014

The Accidental Species: Misunderstandings of Human Evolution by Henry Gee, University of Chicago Press, 2013

ALIENS

What Does a Martian Look Like? The Science of Extraterrestrial Life Jack Cohen and Ian Stewart, Ebury Press, 2004

Bird Brain by Nathan Emery, Ivy Press, 2016

MESSAGES

The Information by James Gleick, Fourth Estate, 2012

From Cells to Civilizations: The Principles of Change That Shape Life by Enrico Coen, Princeton University Press, 2015

Acknowledgements

Firstly, I would like to thank Dan Clifton, who planted the seed that eventually blossomed onto this book. Dan wrote and directed my BBC Horizon documentary *One Degree* and is one of those multi-talented, cross-disciplinary bods who you feel might just be an alien himself. I am hoping that the publication of this book means he will finally stop sending me web links about extra-terrestrials. Just as Dan planted the seed, I am equally grateful to the green fingers of Elly James and Celia Hayley at hhb for training its wandering tendrils and to my radiantly brilliant publisher Antonia Hodgson for encouraging it to unimaginable heights.

Praise be to three gifted researchers: Suzy McClintock, Lizzie Crouch, and Andrew Bailey tirelessly sought the most wonderful stories, the most eminent contributors, and explained bits of biology I am utterly ignorant about with the most other-worldly patience. Thanks also to the crack team at Little, Brown including – but by no means limited to – Rhiannon Smith, Kirsteen Astor, Rachel Wilkie, Sean Garrehy, Zoe Gullen, and everyone on the sales team, especially Jennifer Wilson, Sara Talbot, Rachael Hum and Ben Goddard. I also wish to express my continued gratitude to the legendary Heather Holden-Brown and legistic Jack Munnelly at hhb.

Writing this book has been a great adventure, and I have met some outrageously gifted scientists along the way. Particular mention must go to Mazlan Othman, who was truly inspiring in her enthusiasm for all things alien. Nick Lane was kind enough to give me a basic lesson in biology: a bit like asking Einstein to fix your bike. Nicky Clayton and Nathan Lane not only opened their laboratory, but unlocked their thoughts on the potential biology of intelligent aliens. Warm thanks too to Laurance Doyle, who lucidly explained Shannon's orders of entropy, to Carlos Frenck for turning me on to GRBs, to Jocelyn Bell Burnell for patiently re-telling the extraordinary story of her discovery of quasars, and to Richard Crowther, who gave me the lowdown on Near Earth Objects and the best freebee of the entire book: a model of the *Skylon C1*. And special mention must go to John Elliott, who explained the challenges of decoding alien signals.

It goes without saying that the mistakes within these pages are mine, but there would have been a great deal more of them without the help of my embarrassingly over-qualified referees. Professor Paul Alexander, Head of Astrophysics at Cambridge University was kind enough to correct some howlers in my cosmology, and Dr Ben Slater of the Department of Palaeobiology at Cambridge University put me right on everything from the habitability of red dwarf stars to the deep history of Earth. Professor Peter McClintock of Lancaster University brushed up my thermodynamics, while Dr Nick Lane of University College, London vetted my evolutionary biochemistry without feeling the need to point out that most of it was cribbed from his books in the first place.

ACKNOWLEDGEMENTS

The only thing more daunting than writing a book on a subject you know very little about is being married to someone who is writing a book on a subject they know very little about: thank you so much, Jess, for your love and encouragement. Sonny, Harrison and Lana: this book is for you. Heartfelt thanks, as always, to my ever-supportive family: my mum, Marion, and my sisters Bronwen and Leah, my brothers-in-law Richard, Phil, and Josh, and my in-laws Stephanie and Alan Parker. They, together with friends Alexander and Hannah Armstrong, Torjus and Amelia Baalack, Pierre and Kathy Condou, Steven Cree and Kahleen Crawford, Jono and Amanda Irby, Gary and Lauren Kemp, Bruce McKay and Jonathan and Shebah Yeo have listened patiently to more guff about aliens than anyone would think humanly possible.

Finally I would like to thank Rafael Agrizzi L De Medeiros, Daniella Jackson, Sophie Lewis, and Elouise Ody at my favourite coffee shop, where most of this book was written. I promise to write the next one somewhere else.

Picture Credits